RETURN
LIBRARY
THE STANDARD OIL CO. (OHIO)
WARRENSVILLE L...

RETURN
LIBRARY
THE STANDARD OIL CO. (OHIO)
WARRENSVILLE L...

AUG 0 3 87

Fourier Transform Infrared Spectroscopy

Fourier Transform Infrared Spectroscopy

Industrial Chemical and Biochemical Applications

Edited by

T. Theophanides

*University of Montreal, Department of Chemistry,
Montreal, Quebec, Canada*

D. REIDEL PUBLISHING COMPANY

A MEMBER OF THE KLUWER　　ACADEMIC PUBLISHERS GROUP

DORDRECHT / BOSTON / LANCASTER

Library of Congress Cataloging in Publication Data
Main entry under title:

Fourier transform infrared spectroscopy.

Papers presented at the FT-IR Workshop held under the auspices of the Spectroscopy Society of Canada, Sept. 30–Oct. 1, 1982 in St.-Jovite, Québec, Canada.
Includes index.
1. Infrared spectroscopy–Congresses.   2. Fourier transform spectroscopy–Congresses.   I. Theophanides, Theo M.   II. FT-IR Workshop (1982 : Saint-Jovite, Québec)   III. Société de spectroscopie du Canada.
QD96.I5F69   1984   543'.08583   84-13458
ISBN 90-277-1783-4

Published by D. Reidel Publishing Company,
P.O. Box 17, 3300 AA Dordrecht, Holland.

Sold and distributed in the U.S.A. and Canada
by Kluwer Academic Publishers.
190 Old Derby Street, Hingham, MA 02043, U.S.A.

In all other countries, sold and distributed
by the Kluwer Academic Publishers Group,
P.O. Box 322, 3300 AH Dordrecht, Holland.

All Rights Reserved
© 1984 by D. Reidel Publishing Company, Dordrecht, Holland
No part of the material protected by this copyright notice may be reproduced or
utilized in any form or by any means, electronic or mechanical
including photocopying, recording or by any information storage and
retrieval system, without written permission from the copyright owner

Printed in The Netherlands

TABLE OF CONTENTS

Preface   vii

MOLECULAR VIBRATIONS

A. J. P. ALIX / Theory of Molecular Vibrations. A Survey and
Applications to Absolute and Integrated Infrared Intensities   3

S. F. A. KETTLE / The Intensities of Infrared Spectral Bands   17

EXPERIMENTAL AND INSTRUMENTATION

W. M. GRIMM III, W. G. FATELEY, and J. G. GRASSELLI /
Introduction to Dispersive and Interferometric Infrared
Spectroscopy   25

H. BUIJS / Advances in Instrumentation   43

METAL-MOLECULE INTERACTIONS WITH CHEMICAL AND BIOLOGICAL APPLICATIONS

J. G. GRASSELLI / Industrial Applications of Fourier Transform
Infrared Spectroscopy   55

I. S. BUTLER, J. SEDMAN, and A. A. ISMAIL / FT-IR Spectra of
Coordination Compounds   83

S. F. A. KETTLE / Metal Clusters as Models of Metal Surfaces -
Some Vibrational Aspects   97

T. THEOPHANIDES / FT-IR Spectra of Nucleic Acids and the Effect of
Metal Ions   105

H. MANTSCH / The Application of Fourier Transform Infrared
Spectroscopy to the Study of Membranes   125

ACCESORIES AND DATA PROCESSING TECHNIQUES

K. KRISHNAN / Different Accessories, Main Applications and
Handling Techniques   139

D. G. CAMERON / Data Processing Techniques   159

J. P. COATES and ROBERT W. HANNAH / Computer Based Infrared
Search Systems   167

Author Index   187

Subject Index   189

PREFACE

This volume is a collection of contributions to the FT-IR Workshop held under the auspices of the Spectroscopy Society of Canada and organized by Professor Theophile Theophanides, Director of the Workshop. The gathering of leading spectroscopists and researchers at Gray Rocks to discuss Fourier Transform Infrared Spectroscopy was the occasion of the 29th Annual Conference of the Spectroscopy Society of Canada. The pleasant surroundings of Gray Rocks, St-Jovite, Québec, Canada contributed most positively to the success of the two-day Workshop held September 30, October 1, 1982. The preliminary program and the proceedings were distributed at the Workshop by Multiscience Publications Ltd.

The publication of this volume provides the occasion to thank all the contributors for kindly accepting to lecture at the Workshop and for their collaboration. I thank Mr. Al. Dufresne for accepting to act as manager of the Workshop and Mrs. Susane Dufresne secretary of the Workshop for patiently contacting all the participants and for making the necessary arrangements of registration and accomodation.

THEOPHILE THEOPHANIDES, Professor
Université de Montréal
Department of chemistry
C.P. 6210, Montreal, Quebec
CANADA   H3C 3V1

MOLECULAR VIBRATIONS

THEORY of MOLECULAR VIBRATIONS
A SURVEY and APPLICATIONS to ABSOLUTE and INTEGRATED INFRARED INTENSITIES

Alain J. P. ALIX

(Faculté des Sciences, Université de Reims, B.P. 347, 51062 REIMS, FRANCE)

SUMMARY/INTRODUCTION

A brief outline of the different sets of coordinates employed in the vibrational analysis of the various molecular parameters, and the corresponding transformations are presented in order to give explicit relationships between these different parameters, such as the force constants, the compliants, the Coriolis coupling constants, the mean square amplitudes of vibration ...
The fundamental importance of the $\underline{L}$-matrix is pointed out, as that matrix is linked to all molecular constants.
The definitions and properties of the absolute and integrated infrared intensities are discussed in detail within a matrix formalism which has the same pattern as the conventional ones used to describe the other parameters.
Hence, we give here
(i)   the definitions and the matrix notation of the infrared intensities,
(ii) the relationships between infrared intensities and other parameters,
(iii) their extremal properties (exact and approximate)
(iv) the isotopic intensity shifts under mono- and multisubstitutions, i.e., a set of general (exact) sum rules involving all intensities and one or several isotopes,
(v)   the approximate isotopic intensity shifts derived from application of the first order perturbation theory: these rules are seen to be valid for each individual band separately (additivity rules),
(vi) the infrared intensities corresponding to a completely characteristic mode of vibration.

One has to point out that, for the sake of simplicity and brevity, the basis of the molecular vibration theory is directly presented in a given system of independent symmetry coordinates, and one assumes the following assumptions:
(i)   it exists no redundancy relation linking the internal coordinates,
(ii) the internal and/or symmetry coordinates are rectilinear (the relation between the sets of Cartesian and internal coordinates is linear),
(iii) the harmonic oscillator approximation is used everywhere for keeping a simple matrix notation (quadratic form of the potential energy),
(iv) the Born – Oppenheimer approximation is used to derive the isotopic sum and/or product rules (invariance of the quadratic force constants and of the dipole moment derivatives corrected from rotational contributions).

CONTENTS

I) THEORY OF MOLECULAR VIBRATIONS : a SURVEY

1) Vibrational coordinates and transformations

2) Kinetic energy and potential energy

3) Secular – determinantal equations

4) Transformation relation

5) Molecular constants and parameters : matrix notation

II) INFRARED INTENSITIES

1) Definitions
2) Matrix notation
   a) absolute intensities
   b) integrated absorption intensities
   c) rotational contribution to the dipole moment derivatives
3) Relationships between infrared intensities and other parameters
4) Extremal properties of infrared intensities
   a) exact : dipole moment derivatives
   b) approximate : absolute and integrated intensities
5) Isotopic variation of infrared intensities
   a) general exact sum rules
      (i) involving one isotope and all intensities
      (ii) involving several isotopes and all intensities
   b) approximate rules
      (i) sum and product rules involving one isotope
      (ii) sum rule involving several isotopes and only one intensity band (additivity rule)
6) Infrared intensities corresponding to a completely characteristic mode of vibration

III) REFERENCES

*The references are not numbered in the text. They are classified at the end, according to the subjects.

I) THEORY OF MOLECULAR VIBRATIONS : a SURVEY

1) Vibrational coordinates and transformations

$\underline{X}$: column matrix of Cartesian coordinates
$\underline{S}$: column matrix of independent symmetry coordinates
$\underline{Q}$: column matrix of vibrational normal coordinates

$$\underline{X} = \underline{A}\,\underline{S} \qquad \underline{X} = \underline{L}^X \underline{Q} \qquad (\underline{A} = \underline{M}^{-1}\underline{B}^t\underline{G}^{-1})$$

$$\underline{S} = \underline{B}\,\underline{X} \qquad \underline{S} = \underline{L}\,\underline{Q} \qquad \text{"Fletcher - Powell relation"}$$

2) Kinetic energy and potential energy

$$2T = \underline{\dot{X}}^t \underline{M}\, \underline{\dot{X}} = \underline{\dot{S}}^t \underline{G}^{-1} \underline{\dot{S}} = \underline{\dot{Q}}^t \underline{E}\, \underline{\dot{Q}} \qquad \text{(kinetic energy)}$$

$$2V = \underline{X}^t \underline{f}^X \underline{X} = \underline{S}^t \underline{F}\, \underline{S} = \underline{Q}^t \underline{\Lambda}\, \underline{Q} \qquad \text{(quadratic potential energy)}$$

where $\underline{M}$ is a matrix containing the masses of the atoms, $\underline{E}$ is a unit matrix, and

$$\underline{G} = \underline{B}\,\underline{M}^{-1}\underline{B}^t \qquad \underline{G} = \underline{L}\,\underline{L}^t \qquad \underline{\Lambda} = \underline{L}^t \underline{F}\,\underline{L}$$

$$\underline{G}^{-1} = \underline{A}^t \underline{M}\,\underline{A} \qquad \underline{G}^{-1} = (\underline{L}^{-1})^t \underline{L}^{-1} \qquad \underline{F} = (\underline{L}^{-1})^t \underline{\Lambda}\,\underline{L}^{-1}$$

3) Secular – determinantal equations : G – F Wilson formalism

The description of the motions of the atoms, during the molecular vibrations, is done by using the equations of Lagrange which, in particular, give the fundamental result

$$(\underline{G}\ \underline{F})\ \underline{L} = \underline{L}\ \underline{\Lambda} \qquad \text{and/or} \qquad |\underline{G}\ \underline{F} - \lambda\ \underline{E}| = 0$$

The $\underline{L}$ - matrix appears clearly to be the eigenvector matrix of the matrix product $(\underline{GF})$.
From the invariance of the force constant matrix $\underline{F}$, under isotopic substitution, one gets the Teller – Redlich product rule

$$\prod_i \frac{\lambda_i^*}{\lambda_i} = \frac{|\underline{G}^*|}{|\underline{G}|} = \prod_i \frac{(\omega_i^*)^2}{(\omega_i)^2} \qquad \lambda_i(\text{mdyn/Å amu}^{-1}) = (\frac{\omega_i(\text{cm}^{-1})}{1303.16})^2$$

where the $*$ stands for the isotope.

4) Transformation relations

In the Cartesian system of coordinates, one shows that

$$\underline{L}^X = \underline{M}^{-1}\ \underline{B}^t\ (\underline{L}^{-1})^t \qquad (\underline{L}^X)^t\ \underline{M}\ \underline{L}^X = \underline{E}$$

The above relations are used for the derivation of the variation of the infrared intensity under isotopic substitution. The explicit dependence of the infrared intensities, with respect to the masses of the atoms implicated in the substitution will follow from the above second relation developed in the form

$$\sum_\alpha [(L^x_{\alpha i})^2 + (L^y_{\alpha i})^2 + (L^z_{\alpha i})^2]\ m_\alpha = 1$$

5) Molecular constants and parameters : matrix notation

Force constants: $\underline{F}$ $\qquad \underline{L}^t\ \underline{F}\ \underline{L} = \underline{\Lambda} \qquad\qquad \underline{F} = (\underline{L}^{-1})^t\ \underline{\Lambda}\ \underline{L}^{-1}$

Kinetic constants: $\underline{G}^{-1} \qquad \underline{L}^t\ \underline{G}^{-1}\ \underline{L} = \underline{E} \qquad\qquad \underline{G} = \underline{L}\ \underline{L}^t$

Compliants: $\underline{C} \qquad\qquad \underline{C} = \underline{F}^{-1} \qquad\qquad \underline{C} = \underline{L}\ \underline{\sigma}\ \underline{L}^t \qquad (\sigma = \underline{\Lambda}^{-1})$

Coriolis coupling constants: $\underline{\zeta}^\alpha \qquad\qquad\qquad \underline{\zeta}^\alpha = \underline{L}^{-1}\ \underline{c}^\alpha\ (\underline{L}^{-1})^t$

Mean square amplitudes
of thermal vibrations: $\underline{\Sigma} \qquad \Sigma = <\underline{ss}^t> \qquad\qquad \underline{\Sigma} = \underline{L}\ \underline{\delta}\ \underline{L}^t$
(the thermal averages
are evaluated using
statistical thermodynamics) $\delta_i = <Q_i^2> \qquad\qquad \delta_i = \frac{h}{8\pi^2 c\ \omega_i}\ \coth[\frac{hc\ \omega_i}{2kT}]$

Infrared intensities: (See below)

absolute intensity: $\Gamma_i \qquad\qquad\qquad \underline{\Gamma} = \underline{K}^I\ \underline{L}^t\ \underline{\mu}'\ \underline{L}$

integrated intensity: $A_i \qquad\qquad\qquad \underline{A} = k^{-1}\ \underline{L}^t\ \underline{\mu}'\ \underline{L}$

II) INFRARED INTENSITIES

1) Definitions

The integrated absorption intensity A_i is defined as

$$A_i = \int_\nu k(\nu)\, d\nu \qquad k(\nu) = (1/C\ell)\, \mathrm{Ln}(I_o/I) \qquad \text{(the integration covers the finite width of the band)}$$

C: molar concentration
ℓ: path length of the sample
I_o: intensity of the incident light
I: intensity of the light leaving the sample

Considering the case of nondegenerate fundamental ν_i, one deduces from quantum mechanical derivations (e.g., Einstein's coefficients, partition function, Placzek's approximation for the dipole moment ...), that

$$A_i = N_o\, \pi/3c \sum_\alpha (\partial \mu_\alpha/\partial Q_i)^2 \quad ; \quad \underline{\mu}: \text{dipole moment}, \; Q_i: \text{normal coordinate}$$

For degenerate fundamentals, and when the frequency ν_i is replaced by $\bar{\nu}_i$ (in wave numbers), the general result is

$$A_i = (N_o\, \pi\, d_i/3c^2) \sum_\alpha (\partial \mu_\alpha/\partial Q_i)^2 \quad ; \quad d_i: \text{degeneracy of the i'th fundamental}$$

which is valid only when there is no rotational contribution to the vibrational intensities (i.e., the molecule has no permanent dipole moment or, the irreducible representation does not contain any component of rotation).
In general, infrared intensities contain contributions from vibrational and rotational motions and are governed by additional correction factors.

2) Matrix notation

  a) <u>absolute infrared intensities</u>

$$\Gamma_i = \frac{N_o\, \pi\, d_i}{3c^2} \frac{1}{\omega_i} \sum_\alpha (\partial \mu_\alpha/\partial Q_i)^2 = K_i^I\, (\underline{L}^t\, \underline{\mu}'\, \underline{L})_{ii} \rightarrow \underline{\Gamma} = \underline{K}^I\, (\underline{L}^t\, \underline{\mu}'\, \underline{L})$$

where from $\underline{S} = \underline{L}\, \underline{Q}$, one derives that $(\partial \underline{\mu}/\partial \underline{Q}) = \underline{L}^t\, (\partial \underline{\mu}/\partial \underline{S})$ and where the matrix $\underline{\mu}'$ is defined by

$$(\underline{\mu}')_{ij} = (\partial \underline{\mu}/\partial S_i)(\partial \underline{\mu}/\partial S_j) \quad ; \quad \text{(note that } (\underline{\mu}')_{ii} = (\partial \underline{\mu}/\partial S_i)^2\text{)}$$

  b) <u>integrated absorption intensities</u>

$$A_i = \Gamma_i\, \nu_i = \Gamma_i\, \omega_i \quad \text{(usual approximation)}$$

$$A_i = k^{-1}\, (\underline{L}^t\, \underline{\mu}'\, \underline{L}) \; ; \; (k^{-1}, k \text{ are coefficients}) \qquad \rightarrow \underline{A} = k^{-1}\, (\underline{L}^t\, \underline{\mu}'\, \underline{L})$$

The properties (mathematical) of the matrices $\underline{\Gamma}$ and/or $\underline{A}$ have been discussed in detail (see References).
It may be pointed out that only the diagonal elements of the matrices $\underline{\Gamma}$ and $\underline{A}$ are "observable" physical parameters.
One notes also that the quantities $(\partial \underline{\mu}/\partial S_i)$ are invariant under isotopic substitution (Born-Oppenheimer approximation), provided no vibrational angular momentum results ($\rightarrow \underline{\Delta\mu'} = \underline{0}$).

c) rotational contribution to the dipole moment derivatives

The Eckart conditions indicate that the symmetry coordinates conserve linear and angular momentum. Thus, the vibrational angular momentum causes compensating rotation of the molecule and this, in general, leads to the change of the dipole moment. (See References for some detail). Here, it will be only considered the infrared intensities "corrected" from rotational contribution.

It is not the aim of that lecture to speak about the different interpretations of infrared intensities. So, the "reader" is invited to consult appropriate references about the polar parameters such as:
- $(\partial\mu/\partial S)$,
- polar tensors involving $(\partial\mu/\partial X)$,
- $(\partial e_\alpha/\partial Q_i)$; (bond moment theory → valence optical theory)
- bond charge parameters, ...

3) Relationships between infrared intensities and other molecular parameters

Using the above presented matrix notation, it is easy to derive relations between absolute and/or integrated infrared intensities and other molecular constants such as:
- inverse kinetic constants and force constants: $\underline{G}$, $\underline{F}$
- compliants : $\underline{C}$
- Coriolis coupling constants : $\underline{\zeta}^\alpha$
- mean square amplitudes of vibration : $\underline{\Sigma}$
- quartic centrifugal distortion constants : $\underline{\Theta}$, $\underline{\Phi}$; $\underline{\Theta} = \underline{\Phi}^{-1} = \underline{G}^{-1}\underline{C}\,\underline{G}^{-1}$

From the basic relation $\underline{A} = k^{-1} \underline{L}^t \underline{\mu}' \underline{L} \leftrightarrow \underline{\mu}' = k\,(\underline{L}^{-1})^t \underline{A}\,\underline{L}^{-1}$, one gets

$$\underline{\mu}'\,\underline{G} = k\,(\underline{L}^{-1})^t \underline{A}\,\underline{L}^t \;\rightarrow\; \mathrm{Tr}[\underline{\mu}'\,\underline{G}] = k\,\mathrm{Tr}[\underline{A}] \quad \rightarrow\; \underline{G}\text{-sum rule}$$

$$\underline{\mu}'\,\underline{C} = k\,(\underline{L}^{-1})^t \underline{A}\,\underline{\sigma}\,\underline{L}^t \;\rightarrow\; \mathrm{Tr}[\underline{\mu}'\,\underline{C}] = k\,\mathrm{Tr}[\underline{A}\,\underline{\sigma}] \;\rightarrow\; \underline{F}\text{-sum rule}$$

These results are equivalent to the well-known Crawford's rule. Similar rules can easily be derived for the absolute intensities (Γ).
The generalization of the above Trace-type (sum rule) relations is easily achieved through the following derivation:

for any pair of matrices $\underline{X}$, $\underline{Y}$ satisfying $\underline{X} = (\underline{L}^{-1})^t \underline{Y}\,\underline{L}^{-1}$, one gets

$$\mathrm{Tr}[\underline{\mu}'\,\underline{G}\,\underline{X}] = \mathrm{Tr}[(\underline{K}^I)^{-1}\,\underline{\Gamma}\,\underline{Y}] = k\,\mathrm{Tr}[\underline{A}\,\underline{Y}] \quad \rightarrow\; \underline{G}\text{-sum rules}$$

$$\mathrm{Tr}[\underline{\mu}'\,\underline{C}\,\underline{X}] = \mathrm{Tr}[(\underline{K}^I)^{-1}\,\underline{\Gamma}\,\underline{\sigma}\,\underline{Y}] = k\,\mathrm{Tr}[\underline{A}\,\underline{\sigma}\,\underline{Y}] \quad \rightarrow\; \underline{F}\text{-sum rules}$$

See the Table given below for applications.

$\underline{G}$ - type sum rule		$\underline{Y}$	$\underline{X}$
(G1)	$\mathrm{Tr}[\underline{\mu}'\ \underline{G}] = k\ \mathrm{Tr}[\underline{A}]$	$\underline{E}$	$\underline{E}$
(G2)	$\mathrm{Tr}[\underline{\mu}'\ \underline{\Sigma}] = k\ \mathrm{Tr}[\underline{A}\ \underline{\delta}]$	$\underline{\delta}$	$\underline{G}^{-1}\ \underline{\Sigma}$
(G3)	$\mathrm{Tr}[\underline{\mu}'\ \underline{C}^{\alpha}] = k\ \mathrm{Tr}[\underline{A}\ \underline{\zeta}^{\alpha}]$	$\underline{\zeta}^{\alpha}$	$\underline{G}^{-1}\ \underline{C}^{\alpha}$
(G4)	$\mathrm{Tr}[(\underline{\mu}'\ \underline{G})\ (\underline{F}\ \underline{G})^{m}] = k\ \mathrm{Tr}[\underline{A}\ \underline{\Lambda}^{m}]$	$\underline{\Lambda}^{m}$	$\underline{G}^{-1}\ (\underline{G}\ \underline{F})^{m}\ \underline{G}$
(G5)	$\mathrm{Tr}[\underline{\mu}'\ (\underline{F}\ \underline{G})] = k\ \mathrm{Tr}[\underline{A}\ \underline{\Lambda}]$	$\underline{\Lambda}$	$\underline{G}^{-1}\ \underline{\Phi}$
(G6)	$\mathrm{Tr}[\underline{\mu}'\ (\underline{G}\ \Theta\ \underline{G})] = k\ \mathrm{Tr}[\underline{A}\ \underline{\sigma}]$	$\underline{\sigma}$	$\underline{G}^{-1}\ (\underline{G}\ \Theta\ \underline{G})$

$\underline{F}$ - type sum rule		$\underline{Y}$	$\underline{X}$
(F1)	$\mathrm{Tr}[\underline{\mu}'\ \underline{C}] = k\ \mathrm{Tr}[\underline{A}\ \underline{\sigma}]$	$\underline{E}$	$\underline{E}$
(F2)	$\mathrm{Tr}[(\underline{\mu}'\ \underline{C})\ (\underline{G}^{-1}\ \underline{\Sigma})] = k\ \mathrm{Tr}[\underline{A}\ \underline{\sigma}\ \underline{\delta}]$	$\underline{\delta}$	$\underline{G}^{-1}\ \underline{\Sigma}$
(F3)	$\mathrm{Tr}[(\underline{\mu}'\ \underline{C})\ (\underline{G}^{-1}\ \underline{C}^{\alpha})] = k\ \mathrm{Tr}[\underline{A}\ \underline{\sigma}\ \underline{\zeta}^{\alpha}]$	$\underline{\zeta}^{\alpha}$	$\underline{G}^{-1}\ \underline{C}^{\alpha}$
(F4)	$\mathrm{Tr}[(\underline{\mu}'\ C)\ (\underline{F}\ \underline{G})^{m}] = k\ \mathrm{Tr}[\underline{A}\ \underline{\sigma}\ \underline{\Lambda}^{m}]$	$\underline{\Lambda}^{m}$	$\underline{G}^{-1}\ (\underline{G}\ \underline{F})^{m}\ \underline{G}$
(F5)	$\mathrm{Tr}[(\underline{\mu}'\ \underline{C})\ (\underline{G}^{-1}\ \underline{\Phi})] = k\ \mathrm{Tr}[\underline{A}]$	$\underline{\Lambda}$	$\underline{G}^{-1}\ \underline{\Phi}$
(F6)	$\mathrm{Tr}[(\underline{\mu}'\ \underline{C})\ (\underline{G}^{-1}\ \underline{C})] = k\ \mathrm{Tr}[\underline{A}\ \underline{\sigma}^{2}]$	$\underline{\sigma}$	$\underline{G}^{-1}\ (\underline{G}\ \Theta\ \underline{G})$

The obtained relations serve mainly as a check on numerical calculations. One may note that no relation involving $\underline{\mu}'$ alone could be derived, which could explain the difficulties encountered by some authors (see Escribano), in searching for such relationship between $(\partial\mu_{\alpha}/\partial S_i)^2$ and other parameters.

The choice of dipole moment derivatives is one of the most difficult and crucial problem in infrared intensity theory. Only $(\partial\mu_{\alpha}/\partial Q_i)^2$ are fixed by the experimental intensities. Considering any irreducible representation, containing n infrared active modes, the number of $(\partial\mu/\partial Q)$ -matrices is 2^n which differ only in signs of the different elements (and not in absolute value). The multiplicity of solutions can be reduced by considering the isotopic invariance property (within the Born-Oppenheimer approximation), of the dipole moment derivatives with respect to Cartesian projection of bonds.

A numerical method of fixing a unique set of values for $(\partial\mu_{\alpha}/\partial S_i)$, consists of measuring the intensities corresponding to different isotopic species. In such procedure, the utility of numerical data for different isotopes depends on the sensitivity of the variations with respect to the isotopic substitution. (See the section on isotopic variation of infrared intensities).

4) Extremal properties of infrared intensities

a) exact: dipole moment derivatives

The "exact" extremal and/or stationary properties of the functions involving the derivatives of the dipole moment, $(\partial\mu/\partial S_i)^2$ and $\sum_i (\partial\mu/\partial S_i)^2$, are obtained by the use of the mathematical technique of Lagrangian multipliers.

The term "exact" means that the reported results are based on a rigorous derivation and are strictly valid within the approximations of the harmonic oscillator and the semi-rigid rotor.

(i) $(\partial \underline{\mu}/\partial s_i)^2 = \underline{\mu}'_{ii}$

One obtains the simple result

$(\partial \underline{\mu}/\partial s_i)^2_{stationary} = (\underline{\mu}'_{ii})_{stat.} = (\underline{G}^{-1})_{ii} \sum_k (\partial \underline{\mu}/\partial Q_k)^2 = (\underline{G}^{-1})_{ii} \, Tr[\underline{A}] \, k$

Moreover, it is seen that the force constant associated with the i'th mode of vibration, is fully determined by the constraint $(\underline{\mu}'_{ii})$ = extremal: viz.,

$[F_{ii}] = (\underline{G}^{-1})_{ii} \sum_k [\lambda_k (\partial \underline{\mu}/\partial Q_k)^2]/[\sum_k (\partial \underline{\mu}/\partial Q_k)^2]$

The Table given below illustrates the results for the PF_3 molecule.

ν_k (cm^{-1})	λ_k (mdyn/Å amu^{-1})	Γ_k (cm^2/mM)	$(\partial \underline{\mu}/\partial Q_k)^{exp.}$ (Damu$^{-1/2}$/Å)	$(\partial \underline{\mu}/\partial s_i)^{exp.}$ (D/Å)
$\nu_1(A) = 892$	0.4685	12.360	+ 1.614	+ 4.19 ± 0.89
$\nu_2(A) = 487$	0.1396	4.947	− 0.757	− 2.02 ± 0.05
$\nu_1(E) = 860$	0.4355	48.379	− 2.219	− 7.01 ± 0.30
$\nu_2(E) = 344$	0.0696	2.493	− 0.319	− 0.92 ± 0.01

$|\partial \underline{\mu}/\partial s_i|$	maxi.	CNDO/2	$F_{ii}^{mini.}$	$F_{ii}^{exp.}$	$[F_{ii}]$	$F_{ii}^{maxi.}$
i = 1 (A)	7.10	4.88	2.21	6.23 ↔	6.50 ↔	7.44
i = 2 (A)	4.24	3.59	0.79	0.80 ↔	2.32 ↔	2.66
i = 1 (E)	8.01	9.95*	0.88	4.98 ↔	5.49 ↔	5.56
i = 2 (E)	5.97	1.78	0.49	0.50 ↔	3.04 ↔	3.09

(ii) $[\sum_i (\partial \underline{\mu}/\partial s_i)^2]_{stat.} = Tr[\underline{\mu}']$

It has been demonstrated that

$[\sum_i (\partial \underline{\mu}/\mu s_i)^2]_{stat.} = Tr[\underline{\mu}'] = k \, \gamma_r^{-1} \, Tr[\underline{A}]$

where γ_r is any one of the eigenvalues of the $\underline{G}$-matrix. In particular, when the eigenvalues of $\underline{G}$ are written in increasing order, one obtains the maximum and the minimum values of $Tr[\underline{\mu}']$ by putting r = 1 and r = n respectively in the above Equation. These results could be used to control the values estimated from semi-empirical quantum mechanical methods (see above the case of PF_3).

Application to PF_3 (units are $(D/\overset{\circ}{A})^2$). From the Table one notices that one gets

$(\partial\underline{\mu}/\partial S_i)^2_{max.} = 8.01^2 \rightarrow$ CNDO/2 gives $(-9.95)^2 >$ max.(*)

$\sum_i (\partial\underline{\mu}/\partial S_i)^2_{max.} = 71.8 \pm 5.7 \rightarrow$ CNDO/2 gives $(102.2) >$ max. $S_i = S_1(E)$

b) <u>approximate: absolute and integrated intensities</u>

An approximate study of the extremal properties of the isotopic infrared intensity shifts can be carried out using the first order perturbation theory. It can be shown that the $\underline{L}$-matrices for the mother molecule ($\underline{L}$), and for one of its isotopes ($\underline{L}^*$), satisfy the following relation

$$(\underline{L}^{-1} \underline{L}^*) = \text{diagonal} = \underline{d} = (\underline{\Lambda}^*)^{1/2} (\underline{\Lambda})^{-1/2}$$

where $\underline{d}^2$ is the eigenvalue matrix of the matrix product $(\underline{G}^* \underline{G}^{-1})$. Then, one deduces from the matrix notation of the intensities

$(A^*_i/A_i)_{max.}$ = greatest eigenvalue of $(\underline{G}^* \underline{G}^{-1}) = (d^2_r)_{max.}$

$(A^*_i/A_i)_{min.}$ = smallest eigenvalue of $(\underline{G}^* \underline{G}^{-1}) = (d^2_r)_{min.}$

Further, as it has been shown that the isotopic intensity shifts follow the same rule, one gets

$(A^*_i/A_i)_{max.} = (\lambda^*_r/\lambda_r)_{max.} = (\omega^*_r/\omega_r)^2_{max.}$ etc.

Similar results are obtained for the absolute infrared intensities Γ_i.

5) <u>Isotopic variation of infrared intensities</u>

a) <u>general exact sum rules</u>

(i) involving only ONE isotope and ALL intensities

Using the expression of the infrared intensities, written in the symmetry coordinate system, one gets

$$\text{Tr}[\underline{A}\,\underline{\sigma}] = \sum_i A_i/\omega_i^2 = \text{Tr}[\underline{A}^* \underline{\sigma}^*] = \sum_i A^*_i/\omega^*_i$$

which is the well known Crawford $\underline{F}$-sum rule.

Now, by using the expression of the $\underline{L}^X$-matrix in the Cartesian coordinate system (in which the $\underline{G}$-matrix reduces to the matrix $\underline{M}^{-1}$ containing the reciprocal of the masses of the different atoms constituting the molecule), one obtains the general sum rule

$$\sum_i A_i = - \sum_{\alpha=1}^{N} \sum_i (m^*_\alpha/\Delta m_\alpha) \Delta A_i(\alpha)$$

THEORY OF MOLECULAR VIBRATIONS

(ii) involving SEVERAL isotopes and ALL intensities

Denoting by (α_1), (α_2), ..., (α_p) the mono-substitutions of the aoms α_1, α_2, ..., α_p respectively, and by $(\alpha_1, \alpha_2, ..., \alpha_p)$ the muti-substitution of all atoms $\alpha_1, \alpha_2, ..., \alpha_p$, one has the result

$$\sum_i \Delta A_i(\alpha_1) + \sum_i \Delta A_i(\alpha_2) + ... + \sum_i \Delta A_i(\alpha_p) = \sum_i \Delta A_i(\alpha_1, \alpha_2, ..., \alpha_p)$$

The above result was derived from the use of

$$\sum_{a=1}^{p} \underline{\Delta G}^{(a)} = \underline{\Delta G}^{(pp)} \qquad \text{(valid for any } p = 1, 2, ..., N\text{)}$$

where a and pp stand for mono- and multi-substitution respectively.
The obtained above exact sum rule is valid without regard to the symmetry or the geometry of the molecule. In the cases of symmetric substitutions, the rule is valid for each irreducible representation and reads

$$\sum_{a=1}^{p} \sum_i \Delta A_i^{(a)} = \sum_i \Delta A_i^{(pp)} \qquad \text{(valid for any } p = 1, 2, ..., s\text{)}$$

where a (= 1, 2, ..., p) denotes a mono-substitution of one of the p sets of symmetrically equivalent atoms (the two Y-atoms in XY_2-type molecules of C_{2v} symmetry, for instance), and pp denotes the multisubstitution of the p sets of symmetrically equivalent atoms, taken out of the total of s sets.

For instance, considering the isotopic substitutions $X \to X^*$ and $Y \to Y^*$ in a molecule of the type $[X_m Y_n Z_o ...]$, one deduces from the above sum rule

$$\sum_i \Delta A_i(X_m^* Y_n^* Z_o ...) = \sum_i \Delta A_i(X_m^* Y_n Z_o ...) + \sum_i \Delta A_i(X_m Y_n^* Z_o ...)$$

which leads to

$$\sum_i A_i(X_m Y_n Z_o ...) + \sum_i A_i(X_m^* Y_n^* Z_o ...) = \sum_i A_i(X_m^* Y_n Z_o ...) + \sum_i A_i(X_m Y_n^* Z_o ...)$$

b) <u>approximate rules</u>

(i) sum and product rules involving ONE isotope

Using the first order perturbation theory, it is possible to derive simpler sum rules valid for EACH INDIVIDUAL intensity band separately. One shows that

$$A_i = A_i (\Delta \lambda_i / \lambda_i) \quad \to \quad \updownarrow \quad \begin{array}{l} A_i^*/A_i = \lambda_i^*/\lambda_i = (\omega_i^*/\omega_i)^2 \\ \Gamma_i^*/\Gamma_i = (\lambda_i^*/\lambda_i)^{1/2} = \omega_i^*/\omega_i \end{array}$$

The Table given below illustrates the above results.

Molecule	mode of vibration	isotopic substitution	(Γ_i^*/Γ_i) (measured)	(ω_i^*/ω_i) (calculated)
CF_4	ν_{as} (C-F)	$^{12}C \rightarrow ^{13}C$	1.01 ± 0.02	0.97
Benzoic acid	ν_s (C=O)	$^{12}C \rightarrow ^{13}C$	0.98 ± 0.02	0.98
$POCl_3$ in C_6H_{12}	ν (P-O)	$^{16}O \rightarrow ^{18}O$	0.92 ± 0.03	0.96
	ν_s (P-Cl)	$^{16}O \rightarrow ^{18}O$	0.97 ± 0.03	0.99
	δ_s (PCl_2)	$^{16}O \rightarrow ^{18}O$	$(0.82 \pm 0.16?)$	0.99

A simple approximate PRODUCT rule may also be derived for infrared intensities. This rule has the same form as the one of the famous Teller-Redlich product rule, derived for the frequencies of the fundamentals. One has then

$$\prod_i A_i^*/A_i = \prod_i \lambda_i^*/\lambda_i = \prod_i (\omega_i^*/\omega_i)^2 = |\underline{G}^*|/|\underline{G}|$$

$$\prod_i \Gamma_i^*/\Gamma_i = \prod_i \omega_i^*/\omega_i = \prod_i (\lambda_i^*/\lambda_i)^{1/2} = |(\underline{G}^*)^{1/2}|/|\underline{G}^{1/2}| = (|\underline{G}^*|/|\underline{G}|)^{1/2}$$

See the following Table for applications.

Molecules	$\prod_i (\Gamma_i^*/\Gamma_i)_{obs.}$	$|(\underline{G}^*/\underline{G})^{1/2}|$	$\prod_i (A_i^*/A_i)_{obs.}$	$|\underline{G}^*/\underline{G}|$	Species
$C(H/D)_4$	0.48 ± 0.02	0.578	0.27 ± 0.02	0.334	F_2
$Si(H/D)_4$	0.57 ± 0.03	0.530	0.30 ± 0.03	0.281	F_2
$Ge(H/D)_4$	0.51 ± 0.03	0.513	0.26 ± 0.03	0.263	F_2
$C_2(H/D)_4$	$0.54_6 \pm 0.05$	0.535	0.30 ± 0.05	0.286	B_{1u}

The results presented in the two above Tables, demonstrate clearly that the rules hold rather well in the cases of heavy atom substitution (which is a necessary condition for application of the first order perturbation theory) and are not incompatible even for H/D substitutions.

(ii) sum rule involving SEVERAL isotopes and only ONE intensity (additivity rule)

Using first order perturbation theory, it can be shown that the general exact

THEORY OF MOLECULAR VIBRATIONS   13

sum rule breaks into approximate individual sum rules valid for each individual intensity band. Hence, one has

$$\sum_{a=1}^{p} \Delta A_i^{(a)} = \Delta A_i^{(pp)} \quad \text{(valid for } p = 1, 2, \ldots, s\text{)}$$

which is now an "ADDITIVITY RULE". This means that the shifts in the integrated intensity corresponding to any band (say i) for multi-substitution, is simply the SUM of the corresponding quantities pertaining to mono-substitutions involved in the process. Such result can be used as a check on the experimental data.

Finally, it may be noticed that, when asymmetrical substitutions (i.e., substitutions which lower symmetry) are considered, the equations have to be modified to take account the possibility of accidental degeneracy.

6) Infrared intensities corresponding to a completely characteristic normal mode of vibration

A normal coordinate Q_k is said to be completely characteristic of a symmetry coordinate S_i, if there is only one non-zero element L_{ik} in the k'th column of the matrix $\underline{L}^i$ (which links S_i and Q_k through $\underline{S} = \underline{L}\, \underline{Q}$).

This definition means that a single symmetry coordinate can be used in the description of a normal mode of frequency ν_k.

Hence, for a normal mode of frequency ν_k, one has for the general case where no defined assignment has been given a mixing of the symmetry coordinates which can be "measured" by various methods (e.g., PED = potential energy distribution).

In the special case of a completely characteristic normal mode of vibration, one has

$$S_i^{(k)} = L_{ik} Q_k \; ; \quad S_j^{(k)} = 0 \quad (j = 1, 2, \ldots, n, \text{ except } j = i)$$

In the usual "a priori" assignment, one has i = k (i.e., $L_{ii} \neq 0$ for all i). A completely characteristic normal mode of vibration gives as a particular result

$$\partial \underline{\mu}/\partial Q_i = \pm (\underline{G}^{-1})_{ii}^{-1/2} (\partial \underline{\mu}/\partial S_i)$$

So, using the above result in the expression of the integrated infrared intensity, one gets

$$A_i = \frac{N \pi d_i}{3 c^2} \cdot \frac{(\mu')_{ii}}{(\underline{G}^{-1})_{ii}} = \frac{N \pi d_i}{3 c^2} \cdot \frac{(\partial \underline{\mu}/\partial S_i)^2}{(\underline{G}^{-1})_{ii}}$$

Very good values for the bond moment parameters $\partial \underline{\mu}/\partial S_i$ are found by using these expressions.

It is well known that many molecules exhibit characteristic modes of vibration (especially bending modes) and then, the above result may be applied in such cases.

It is also known that certain parameters attain one of their extremal values (sometimes only stationary values) in the case of completely characteristic mode of vibration (e.g., force constant F_{ii}, compliant C_{ii}, mean square amplitudes of vibration Σ_{ii}, ...) but it can been shown that this is not the case for $(\partial \underline{\mu}/\partial S_i)^2$.

A similar study and analogous results are obtained in the case of the definition of a completely characteristic generalized normal force.

Although no experimental verification of ALL the above obtained equations is possible at present, for a lack of data, a comparison with similar results for the other molecular constants and/or parameters would give some confidence that such results could also be accurate (specially for symmetric isotopic substitutions involving heavy atoms).

III) REFERENCES

THEORY OF MOLECULAR VIBRATIONS: a SURVEY

1. E.B. Wilson, Jr., J.C. Decius and P.C. Cross, "Molecular Vibrations", Mc Graw-Hill, New York (1955).
2. S.J. Cyvin, "Molecular Vibrations and Mean Square Amplitudes", Elsevier, Amsterdam (1968).
3. K. Nakamoto, "Infrared Spectra of Inorganic and Coordination Compounds", John Wiley, New York (1970).
4. D. Steele, "Theory of Vibrational Spectroscopy", W.B. Saunders, London (1971).
5. L.A. Woodward, "Introduction to the Theory of Molecular Vibrations and Vibrational Spectroscopy", Oxford University Press, London (1972).

INFRARED INTENSITIES

DEFINITIONS

6. G. Herzberg, "Molecular Spectra and Molecular Structure (Vol. 1, 2, 3), Van Nostrand, New York (1939, 1945, 1966).
7. See Ref. 1.
8. H. Eyring, J. Walter and G.E. Kimball, "Quantum Chemistry", John Wiley, New York (1944).
9. M. Born and K.R. Huang, "Dynamical Theory of Crystal Lattices", Clarendon Press, London (1954).
10. J. Overend, in "Infrared Spectroscopy and Molecular Structure", (Ed. M. Davies), Elsevier, Amsterdam (1963).
11. S. Califano, "Vibrational States", John Wiley, New York (1966).
12. P.W. Atkins, "Molecular Quantum Mechanics", Oxford University Press, London (1970).
13. L.A. Gibov, "Intensity Theory of Vibrational Spectroscopy", W.B. Saunders, London (1971).
14. L.M. Sverdlov, M.A. Kovner and E.P. Krainov, "Vibrational Spectra of Polyatomic Molecules", John Wiley, New York (1974).
15. C.J.H. Schutte, "The Theory of Molecular Spectroscopy", North Holland, Amsterdam (1976).

CORRECTION FACTORS

16. B.L. Crawford, Jr., and H.L. Dinsmore, J. Chem. Phys., 18 (1950) 983 ; ibid., 18 (1950) 1682.

17 R. Herman and R.F. Wallis, J. Chem. Phys., 23 (1955) 637.
18 H.M. Manson and H.H. Nielsen, J. Mol. Spectrosc., 4 (1960) 468.
19 R.A. Toth, R.H. Hunt and E.K. Plyler, J. Mol. Spectrosc., 32 (1969) 74 ; ibib., 32 (1969) 85 ; ibid., 35 (1970) 110.
20 R.H. Tipping, J. Mol. Spectrosc., 61 (1976) 272.

ROTATIONAL CONTRIBUTION TO THE DIPOLE MOMENT DERIVATIVES

21 B.L. Crawford, Jr., J. Chem. Phys., 20 (1952) 977.
22 A.D. Dickson, I.M. Mills and B.L. Crawford, Jr., J. Chem. Phys., 27 (1957) 445.
23 J.W. Russel, C.D. Needham and J. Overend, J. Chem. Phys., 45 (1966) 3383.
24 W.T. King, G.B. Mast and P.P. Blanchette, J. Chem. Phys., 56 (1972) 440.
25 A.J. van Sraten and W.M.A. Smit, J. Mol. Spectrosc., 56 (1975) 484.
26 M. Gussoni and S. Abbate, J. Mol. Spectrosc., 62 (1976) 53.
27 R. Escribano, G del Rio and J.M. Orza, Mol. Phys., 33 (1977) 543.

POLAR PARAMETERS

28 M. Gussoni, G. Dellapiane and S. Abbate, J. Mol. Spectrosc., 57 (1975) 323.

POLAR TENSORS

29 J.F. Biarge, J. Herranz and J. Morcillo, An. R. Soc. Esp. Fis. Quim., 57A (1961) 81.
30 J. Morcillo, L.J. Zamorano and J.M.V. Heredia, Spectrochim. Acta, 22 (1966) 1569.
31 M. Lastra, A. Medina and J. Morcillo, An. R. Soc. Esp. Fis. Quim., 63B (1967) 1079.
32 J. Morcillo, J.F. Biarge, J.M.V. Heredia and A. Medina, J. Mol. Struct., 3 (1969) 77.
33 See Ref. 24
34 W.B. Person, S.K. Rudys and J.H. Newton, J. phys. Chem., 79 (1975) 2525
35 W.B. Person and J.H. Newton, J. Chem. Phys., 61 (1974) 1040.
36 J.H. Newton and W.B. Person, J. Chem. Phys., 64 (1976) 3036.
37 W.B. Person and J. Overend, J. Chem. Phys., 66 (1977) 1442.
38 J.C. Decius, J. Mol. Spectrosc., 57 (1975) 348.

VALENCE OPTICAL THEORY

39 M.V. Volkenshtein and M.A. Elyashevitch, Doklady AN SSSR, 41 (1943) 380 ; ibid., 43 (1944) 56.
40 M.V. Volkenshtein, Doklady AN SSSR, 32 (1941) 185.
41 See Ref. 13.
42 M.V. Volkenshtein, L.A. Gribov, M.A. Elyashevitch and Y.Y. Stepanov, "Kolebanya Molekul", Yzhdatelstvo, Nauka, Moscow (1972).
43 See Ref. 14.

44 L.M. Sverdlov, Opt. Spectrosc., 35 (1973) 37 ; ibid., 39 (1975) 113.
45 M. Gussoni and S. Abbate, J. Chem. Phys., 65 (1976) 3439.

BOND CHARGE PARAMETERS

46 L.S. Mayants and B.S. Averbukh, J. Mol. Spectrosc., 22 (1967) 197.
47 A.J. van Straten and W.M.A. Smit, J. Mol. Spectrosc., 62 (1976) 297 ; ibid., 65 (1977) 202.
48 A.J. van Straten, "A Bond Charge Parameter Theory for Integrated Infrared Intensities", Thesis, Utrecht (1977).

MATRIX NOTATION, EXTREMAL PROPERTIES, ISOTOPIC RULES, CHARACTERICITY

49 See Ref. 21.
50 A.J.P. Alix, N. Mohan, A. Müller and S.N. Rai, Z. Naturforsch., 28a (1973) 1408.
51 N. Mohan and A Müller, J. Mol. Struct., 27 (1975) 255.
52 A.J.P. Alix, C. R. Acad. Sci. Paris, 281B (1975) 477.
53 A.J.P. Alix, J. Mol. Struct., 33 (1976) 137.
54 N. Mohan, A.J.P. Alix and A. Müller, "31st Symposium on Molecular Spectroscopy", Columbus, Ohio, June 14 - 18 (1976) (session M32, speaker A.J.P.A.)
55 N. Mohan, A.J.P. Alix and A. Müller, Mol. Phys., 33 (1977) 319.
56 A.J.P. Alix and N. Mohan, C. R. Acad. Sci. Paris, 285B (1977) 141.
57 A. Müller and N. Mohan, J. Chem. Phys., 67 (1977) 1918.
58 A.J.P. Alix and A. Müller, J. Chim. Phys., 74 (1977) 727.
59 N. Mohan and A. Müller, J. Mol. Spectrosc., 80 (1980) 455.
60 A.J.P. Alix and E. Rytter, Z. Naturforsch., 35a (1980) 1142.

THE INTENSITIES OF INFRARED SPECTRAL BANDS

S.F.A. Kettle
School of Chemical Sciences, University of East Anglia, Norwich
NR4 7TJ, U.K.

In the visible and near ultraviolet region of the spectrum it is customary when listing spectral data to give both the positions of band maxima and also their extinction coefficients. In contrast, in the infrared it is band positions that are quoted with precision, band intensities being reduced to a v.s., s, m, w, v.w (very strong → very weak) classification.

Traditionally, the absolute intensities of infrared spectral bands have been difficult to measure accurately and there is now less activity in the area than was once the case. However, with the advent of F.T. I.R. techniques, the subject seems ripe for a revival.

What has been the historial reason for the interest in accurate infrared band intensities, be they relative or absolute? Several reasons may be listed. Firstly, to supplement frequency data in vibrational analyses. Vibrational analyses on molecules of any size are usually under-determined - even data on isotopomers may be less helpful than one would hope because of frequency shifts which are comparable to frequency errors, the overlapping of bands and ambiguities of interpretation. For molecules lacking a centre of symmetry it frequently happens that, for instance, one symmetry coordinate transforms as z^2, x^2+y^2 or x^2-y^2 - and so, in zeroth order, is almost without infrared intensity - whilst another symmetry coordinate of the same symmetry species transforms as z and so is infrared active in zeroth order (A_1 modes in C_{4v} provide an example of this). The way that the infrared intensity is shared between the two bands in the actual spectrum is, then, indicative of the mixing between the symmetry coordinates and thus provides a further constraint in a normal coordinate analysis.[1]

Secondly, since infrared intensities originate in dipole moment changes - which have vectorial properties - within molecules and, further, these dipole moment changes may, as a first approximation, be associated with internal coordinates such as bond length and bond angle changes there is the possibility of obtaining relatively molecule-independent parameters which, when added vectorially, may be used in predicting infrared band intensities in other molecules.[2]

Thirdly, if the atoms in a molecule carry small residual charges then the movement of these atoms may give rise to a dipole moment change. The amplitude of atomic motions in a vibration may be obtained from force field analyses so these data, together with absolute infrared intensities, may enable residual atom charges to be determined [2,3] assuming that such charges are little changed during the vibrational mode.

Fourthly, when two symmetry-related internal coordinates combine to give several infrared-active bands then, provided that mixing with other internal coordinates can be neglected, the relative intensities of the resulting bands can, by simple vector arguments, be used to obtain the angle between the internal coordinates. Bond angles between CO groups in metal carbonyl derivatives have been determined in this way, usually rather accurately.[4]

Finally, one possible test of the accuracy of detailed calculations on molecular structure, be they ab initio or semi-empirical, is to calculate the absolute intensities of infrared experimental bands.[5]

This list does not exhaust the applications of infrared intensity measurements - for instance, they have been used to obtain data on intermolecular interactions[6,7] - but those described above seem to be the most important.

There are significant sources of error in the measurement of infrared intensities, so that values reported are often regarded as accurate to no more than 5%. For conventional spectrometers the inlet and exit slits represent a problem.[8] Firstly, their finite size means that they sample a range of frequencies; the radiation transmitted is not purely monochromatic. It is usually assumed that the slit function is triangular, peaking at the 'monochromatic' frequency, transmission dropping rapidly, and linearly with frequency separation from this maximum. Secondly, but usually of little importance, for fixed slits the spectral resolution varies over the spectrum. Next, and potentially much more of a problem than might be expected, are the settings of the 0% and 100% absorbance levels, where a setting error of 1% can apparently yield an error of up to 13% in band area.[9] Next, for solution studies, a change in solvent can lead to a variation in band intensities. Evident examples of this are provided by the intensity enhancement observed for hydrogen-bonded species in the liquid phase when compared to the gaseous. With careful choice, this problem can be reduced but the need for solvent 'windows' may well limit the choice available. Variation in choice of solvent can, in bad cases, lead to an order of magnitude change in band intensities.[10] It is not just the central, band maximum, region which contributes to the integrated band intensity. The wings also make an appreciable contribution and a correction has to be applied. This correction becomes both larger and more difficult to evaluate as bands increasingly overlap one another.[11] In the expressions usually quoted which relate the spectral measurements to a vibrational intensity there is a summation over rotational levels which may introduce substantial error, particularly when the rotational line width is much less than the slit width. Finally, and especially for samples well away from room temperature, failure to chop the beam before the sample - or to double chop the beam - introduces an error due to the measurement of superimposed emissions and absorption bands.

Traditionally, actual peak areas have been measured by methods ranging from the simple - cutting out and weighing, through planimetry to digitisation and computer manipulation. Which method has been used seems to depend more on the number of measurements to be made than anything else.[11,12]

How does the advent of F.T.I.R. spectroscopy change this picture? Rather little work has as yet been published on the measurement of absolute infrared intensities by F.T. methods, although the ability of the technique to accurately and rapidly handle spectral subtractions is a major aspect of most of its applications. The Fellgett and Jacquinot advantages - or, rather, the consequent signal to noise improvement - the wavenumber independence of the resolution and the wavenumber accuracy, all represent an advantage of F.T.I.R. techniques. In a recent paper it was commented that

"There is no question that the determination of integrated infrared intensities on an F.T.I.R. spectrometer is much easier and quicker than with conventional dispersive instruments".

Easier and quicker, yes, but is it also more accurate? It is not possible to give an answer to this question but there are hints that if the answer is 'yes' it will be because of the computer facilities associated with F.T.I.R. Thus, Scanlon, Laux and Overend[13] obtained by F.T.I.R. methods an integrated intensity for CO_2 (over the vibration-rotation bands at ca 670 cm^{-1}) which was essentially identical, in value and error, to one obtained twenty years earlier using a grating instrument.[14] Because several of the problems attendant on the measurement of absolute infrared intensities do not apply to gases, and the slit-width problem may be reduced by pressure broadening with an inert gas, the measurements were, in both cases, rather good - an error of just over 1%.

For spectra of solutions, it has been found that a greater accuracy is usually obtained if F.T.I.R. spectra are not apodized.[15] However, for very sharp bands (with band widths smaller than the instrumental resolution) without apodization side-lobe wiggles appear on either side of the band itself. In such cases it is better to apodize with a triangular rather than box-car function. This situation may be compared with the rule-of-thumb for conventional infrared spectrometers that the peak width at half height should be at least five times the instrumental spectral slit width if accurate results are to be obtained.[11]

A useful comparison of F.T. and conventional I.R. instrumentation has been given by Chenery and Sheppard.[16] The advantages of the former - which have yet to be fully exploited in the field of absolute infrared intensities are:

(a) High resolution at a given signal to noise ratio so that the slit-width problems of conventional instruments are either eliminated or reduced. For gas phase spectra, the need to pressure broaden very sharp lines is reduced because the ultimate F.T. resolution may approximate Doppler-broadened line widths.

(b) High signal to noise ratio at a given resolution, leading to more accurate measurements.

(c) Freedom from stray light, particularly for the far infrared. However, fold-back problems must be avoided.

(d) If the sample is placed after the interferometer, sample emission effects are easily eliminated.

Early F.T.I.R. instruments had a poor reputation for intensity measurements, almost certainly because of a combination of source instability and inadequate computational facilities (too few points being used). Both of these two problems have now been eliminated;[16] any residual problem due to long term source instability may be overcome by using a second, dummy, reference.[13]

REFERENCES

1. Recent examples of the way that such data may be used along with a variety of other measurements are provided by G.A. Battiston, G. Bor, U.K. Dietler, S.F.A. Kettle, R. Rossetti, G. Sbrignadello and P.L. Stanghellini, Inorg.Chem., 19 (1980) 1961 and J.C. Decius and D. Murhammer. Spect.Act, 37A (1980) 965.

2. This search has not often proved particularly fruitful. This is one topic discussed in a useful review of infrared intensities by D. Steele, Quart.Rev., 18 (1964) 21. See also D.F. Hornig and D.C. McKean, J.Phys.Chem., 59 (1955) 1133. However, some success has been achieved with the help of more parameters; see L.A. Gribor "Intensity Theory for Infrared Spectra of Polyatomic Molecules" Consultants Bureau, New York, 1964. L.M. Sverdlav, M.A. Kovner and E.P. Krainov, "Vibrational Spectra of Polyatomic Molecules" Wiley, New York, 1974.

3. See, for example, P.R. Davies and W.J. Orville-Thomas, J.Mol.Struct., 4 (1969) 163.

4. S.F.A. Kettle and I. Paul, Adv.Organometallic Chem., 10 (1972) 199.

5. See, for example, W.M.A. Smit and T. van Dam, J.Chem.Phys., 72 (1980) 3658. W.B. Person and J.H. Newton, J.Mol.Struct., 46 (1978) 105. Interesting insights into the problems in this area may be obtained from a series of papers in J.Quant.Spect. Rad.Transfer, 20 (1978) 291; 24 (1980) 75 and 77 and from the references quoted therein.

6. See, for example, J.M. Angellelli, A.R. Katritzky, R.F. Pinzelli and R.D. Topsom, Tetrahedron 28 (1972) 2037.

7. See other, rather different, applications given by N. Mohan, A.J.P. Alix and A. Müller, Mol.Phys., 33 (1977) 319.

8. J.A.J. Thompson, Spect.Acta, 14 (1959) 145. A.L. Khidir and J.C. Decius, Spect.Acta, 18 (1962) 1629.

9. G.J. Boobyer, Spect.Acta, 23A (1967) 335.

10. S. Tanaka, K. Tanabe and H. Kamada, Spect.Acta, 23A (1967) 209.

11. D.A. Ramsay, J.Amer.Chem.Soc., 74 (1952) 72.

12. A. Cabana and G. Sanderby, Spect.Acta, 16 (1960) 335.

13. K. Scanlon, L. Laux and J. Overend, Appl.Spect., 33 (1979) 346.

14. J. Overend, M.J. Youngquist, E.C. Curtis and B. Crawford Jr. J.Chem.Phys., 30 (1959) 532.

15. R.J. Anderson and P.R. Griffiths, Anal.Chem., 47 (1975) 2339.

16. D.H. Chenery and N. Sheppard, Appl.Spect., 32 (1978) 79.

EXPERIMENTAL AND INSTRUMENTATION

INTRODUCTION TO DISPERSIVE AND INTERFEROMETRIC INFRARED SPECTROSCOPY

Wm. M. Grim, III and W. G. Fateley
Department of Chemistry
Kansas State University
Manhattan, Kansas   66506

and

J. G. Grasselli
4440 Warrensville Center Road
Standard Oil Company (Sohio)
Cleveland, Ohio   44128

ABSTRACT

The basic concepts of dispersive and interferometric infrared spectroscopy are dealt with in this chapter. An historical approach is taken in which many of the problems encountered in the development of both techniques are discussed along with the modifications used to solve them. The benefits, drawbacks, and limitations of both techniques are discussed. The materials and instrumentation used in infrared spectroscopy are outlined and discussed, but specific implementation is left to other sources, and sample-handling techniques should be studied in more comprehensive texts on sampling. A short bibliography is included which will allow the investigator to research selected topics more thoroughly.

In this chapter will be presented an introduction to infrared spectroscopy. It is by no means intended to be all-inclusive; whole books which have been written on highly specialized aspects of infrared spectroscopy do not even claim to be exhaustive in their own specialty. It is meant to provide a base for the beginning spectroscopist, and it is also meant to provoke a host of questions. The questions usually will not be difficult to answer; it is merely beyond the scope of one chapter to provide all the answers. A novice does require at least a vague notion of where to begin the quest; it is hoped that this chapter will provide stimuli and point the novice toward the most useful path. Several basic texts on both dispersive and interferometric infrared spectroscopy that have proved to be invaluable to the authors are included as references. Further chapters will also serve to explain more detailed concepts of and applications for infrared spectroscopy.

Infrared (IR) spectroscopy is one of the most versatile, fast, inexpensive, and conclusive methods of molecular identification available to the instrumentally-oriented analytical chemist. Samples can usually be examined with ease and are not limited to one or two phases: gases, liquids, and solids can usually be examined with equal facility, a feature available in few other instrumental techniques. Specialized techniques have been developed to handle all but the most stubborn samples. Some of these techniques are summarized in Table 1, and some are described in other chapters. One drawback to infrared spectroscopy is caused by the properties of water, which can affect the samples, the spectrum, and the instrument. Proper sample handling and instrumental technique can reduce this problem. Once IR spectra are obtained, simple rules can be followed to suggest possible structural features of the unknown compound, and massive spectral libraries have been stored in books and computer files to help to verify tentative identifications.

Table 1

Sample handling Techniques

1. Gases - may be the easiest; windows selected for region of interest 10 cm path length common - instrument will usually accommodate 20 cm cell Folded Path allows for effective path lengths of 1m...10m...20m...

2. Liquids
   a) thin films... between two plates infrared transparent crystals
   b) cavity cells and two infrared crystal plates separated by spaces (0.01, 0.05, .1mm)
   c) attenuated total reflection - ATR
   d) reflection
   e) dissolved solute in a solvent
   f) encapsulation in KBr pellet... reaction studies... quick and dirty techniques when no cell available

3. Solids
   a) Single crystal - transmission, reflection, or emission
   b) ATR - contact of sample to KRS-5 crystal very important
   c) Halide pellet - grinding may lead to polymorphism
   d) Nujol mull - Nujol hands may obscure important sample bands
   e) diamond cells - high pressure 100,000 atm or 100 K bar
   f) dissolve in a solvent, and run as a liquid - some changes in spectral features can be expected
   g) <u>hit it with a hammer</u>...AgCl... pellet also polyethylene - sample for far ir
   h) diffuse reflectance

Infrared spectroscopy is an art and a science that has evolved greatly in the 90 years since W. W. Coblentz began investigating the infrared spectra of "disturbed molecules", mostly rock crystals. After years of study, it was ascertained that the spectra were related to the rotational and vibrational energy state transitions of the molecules. In recent years the technique has diverged into two apparently different methods of spectroscopy: dispersive spectroscopy, and interferometric spectroscopy. In addition, complementary vibrational and rotational spectroscopic techniques are available, including Raman and Electron Energy Loss Spectroscopy. Although the dispersive and interferometric techniques are very different in appearance, and a problem in one technique may not be present in the other technique, many of the same guidelines and problems apply to both types of instrumentation. It is therefore useful to discuss the two techniques in the same context, although not simultaneously. The similarities are thereby dealt with together, and the differences can be highlighted and explained.

Very broadly defined, the difference between dispersive spectrometry and interferometry is a difference of scanning domains. Both instrumental techniques can use the same sources, but the spectra are acquired by different means. A dispersive spectrometer uses a grating or a prism to disperse radiation from the source into spectral elements, and a detector measures the energy of each spectral element. A dispersive spectrometer measures the energy for each spectral element along the frequency domain, and vibrational and rotational bands are observed directly from the spectrum. An interferometer, on the other hand, splits radiation from the source between two optical paths. The radiation from each path is reflected by mirrors and returned along the same paths to recombine, usually before interacting with the sample. The radiation from the two paths can either interfere constructively or destructively depending on the phase difference of the two optical paths, and the resulting pattern is an interferogram, which is a measurement of the energies as a function of optical path difference. The domain in this case is the optical path difference. Interferograms are extremely difficult if not impossible to interpret visually. In order to record a vibrational transition band, the interferogram must be transformed into a spectrum using Fourier Transform mathematics.

The two methods of scanning lead to some inherent differences in the construction of various components of the spectrometer and the interferometer. These differences will be noted briefly here in order that comparisons can be made, and they will be explained further in appropriate sections. Dispersive spectrometry should be thought of as an instrumental technique that is mechanically tolerant and can withstand fairly rough handling, is energy-limited, and has a slow but sensitive detector response. Interferometric techniques provide much more energy, but the nature of the scanning requires detectors with a much faster response rate, and thus far these detectors are inherently less sensitive than those used in dispersive spectroscopy. Interferometry can yield very precise measurements, but the instrument itself requires extremely accurate alignment, and hence it is very mechanically intolerant. Interferometers and their attached computers can also be much more expensive than a survey dispersive instrument, but the price difference between these two techniques is slowly eroding, and many dispersive instruments are now equipped with microcomputers as well. These concepts should be considered in the more detailed explanations.

The first infrared spectrometers to be developed were single-beam dispersive spectrometers. Initially natural rock salt and quartz prisms were used to disperse radiation, but prisms have several limitations, such as limited transmission at certain regions of the spectrum, limited resolution, hygroscopicity, and various distortions for which it is difficult to compensate. Prisms disperse radiation linearly with respect to wavelength, so all spectra obtained using a prism spectrometer are presented with the abscissa linear with respect to wavelength. A typical wavelength range is 2.5 to 25.0 μm. Gratings make improved resolution attainable, but necessitate more complicated optics. With the advent of grating spectrometers, which disperse radiation linearly with respect to frequency, the format of plotted spectra began to change. Actual frequencies are on the order of 10^{14} Hz, and are rather unwieldy values, so the wavenumber $\tilde{\nu}$, defined as ν/c or $1/\lambda$, is used. The tilde is often omitted by spectroscopists, who often use ν. Wavenumber units are expressed in cm^{-1}. A typical wavenumber range is therefore 4000 - 400 cm^{-1}. Special features can be added to extend this range to about 8000 - 100 cm^{-1}. Interferometers are capable of measurements in the region from above 20,000 - 10 cm^{-1}. The near infrared (NIR) region is conventionally defined to be the region from 20,000 - 4,000 cm^{-1}, and the far infrared (FIR) region is conventionally defined to be the region from 200 - 10 cm^{-1}.

Ideally, the components of a spectrometer should either totally reflect or totally transmit infrared radiation, while the surrounding walls should absorb all stray light, and not emit any radiation. Thus, the mirrors and gratings are made as reflective as possible, while prisms and cell walls are composed of materials which are transparent to infrared radiation. Prisms are usually constructed from NaCl, KBr, or CsI, which are very uniformly transparent from over 20,000 cm^{-1} to 650, 400, and 200 cm^{-1} respectively. All IR transmission samples require a substrate on or a cell in which they can be placed. Common cell materials are summarized in Table 2. The ideal cell material would be 100% transparent in the spectral region of interest, non-interactive with the sample, hard but not brittle, inexpensive, non-hygroscopic, and insoluble in water. Needless to say, no materials fulfill all these requirements, so cell materials are chosen to have the properties necessitated by the sample or experiment.

Table 2
Materials used as windows for infrared radiation

		Solubility	Region, cm^{-1}	Usefulness
a.	Quartz	none	10,000-5,000	limited
b.	Silicon	none	5,000-10	limited
c.	Diamond	none	mid ir	expensive
d.	NaCl	yes	mid ir	practical, cheap
e.	KBr	yes	mid ir	practical, cheap
f.	CsI/CsBr	yes	5,000-200	useful, expensive, collects water
g.	AgBr/AgCl	none	5,000-400	very soft, expensive, hard to polish
h.	Polyethylene	none	500-10	inexpensive-limited

There are other materials for special occasions.

A simple single-beam spectrometer is depicted in Figure 1. The spectrometer consists of a source of radiation, a collecting mirror, an entrance slit, a dispersion element, an exit slit, and a detector. Each of these components will be examined separately.

The source for a spectrometer can be any black-body radiator, as has been characterized by Planck, Jeans, and others. When the sample is the source, whether a distant star or a lump of coal, the spectrum of the infrared emission can be measured. Infrared transmission is measured when radiation from a source is directed onto the surface or through a sample.

Fig. 2 illustrates two characteristics of Planck black-body radiation. It can be seen that as a source is heated, the λ_{max} is shifted to higher frequencies, and the intensity at each frequency is increased in a non-linear manner. Intuitively, one would assume that as the source temperature is increased, the sensitivity would also be increased. Two factors counteract the benefits of increased radiation from a hotter source. As the source temperature is increased, the intensity of multiple orders of radiation emanating from the dispersive element is increased disproportionately.

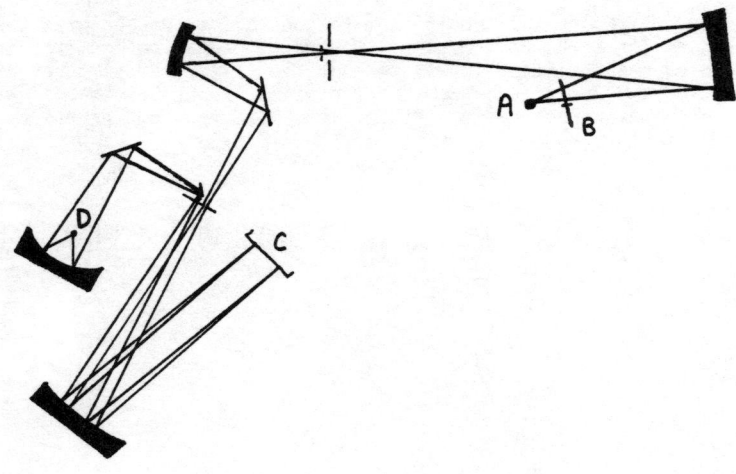

Fig. 1. A. Source; B. Chopper; C. Grating Rack; D. Detector.

Furthermore, very hot sources heat the rest of the spectrometer, and lead to emission witin the spectrometer as well as thermal warping of the optics. Sample heating can also lead to sample emission errors. Thus there are optimum source temperatures which depend on the source, the conditions of the experiment, and the detector. A list of common sources is provided in Table 3. Tunable dye lasers are emerging as very precise sources, with 10^{-6} cm^{-1} resolution often attainable. The wavenumber range of tunable lasers is severely limited thus far, so their applicability has been fairly restricted.

Table 3

"Hot" Objects

Approximate Useful IR-Range

W-Lamp	10,000 to 3,000 cm^{-1}	Near IR
Nernst glower	5,000 to 3,000 cm^{-1}	Near IR
Globar	5,000 to 100 cm^{-1}	Mid and Far IR
Hg-Arc	250 to 10 cm^{-1}	Far IR
Opperman		Mid IR

The collecting mirrors are used to maximize the amount of radiation available to the spectrometer. It should be noted that λ_{max} is not shifted, and the instrument does not suffer from the thermal effects of a hotter source. Typical values for the f-number of the collecting mirrors range from f5.6 to f11.0 for commercial instruments.

The entrance slit serves to limit the radiation entering the spectrometer. Although this would seem to contradict the goal mentioned above, it is done for a reason. Most of the energy emanating from the source is focused so that it will pass through the slit. The slit serves to prevent most of the unwanted sources of radiation, known as stray light, from entering the spectrometer and spuriously increasing the energy measured by the detector.

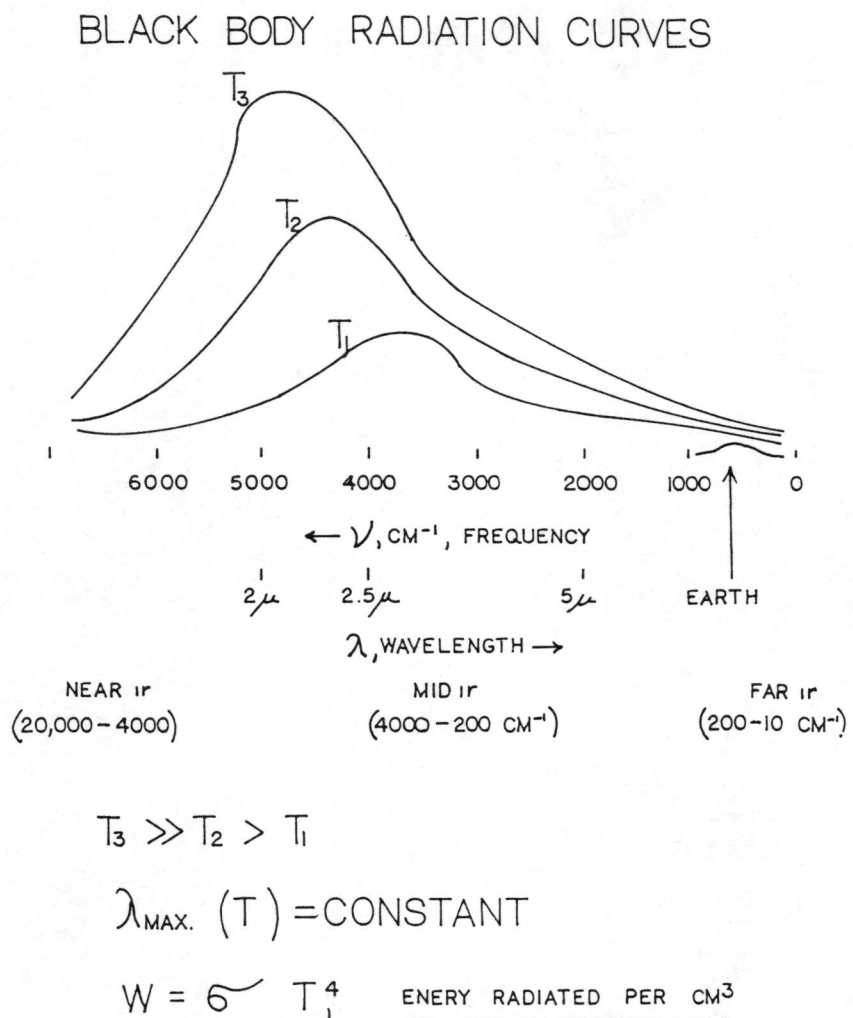

Fig. 2. The Effect of Temperature Elevation on the Graph of Planck Black-body Radiation.

The grating or the prism is the element in the spectrometer that disperses the radiation into spectral elements. The dispersion equation is given by

$$n\lambda = 2d\sin\theta$$

where

n = order
λ = wavelength
d = # lines etched per mm of grating
θ = grating movement

It can easily be seen from this equation that radiation from a second order of wavelength $\lambda/2$ or a third order of wavelength $\lambda/3$ will be registered as energy on the detector at λ. Similarly, radiation from second and third orders of λ will register at 2λ and 3λ respectively. This is known as false energy, and should not be confused with overtone or combination bands, which are actual harmonic oscillations. Filters are therefore needed to screen out the higher order bands of lower wavelength radiation. Lower wavelengths correspond to higher wavenumbers, so in the example illustrated in Figure 3, radiation from the 2nd order of a band at 2250 cm^{-1}, will be detected at 1125 cm^{-1}, and radiation from the 3rd order will be detected at 750 cm^{-1}. Several modern grating instruments can scan from 4000 to 2000 cm^{-1} using only two gratings and two filters. Interferometric techniques do not use gratings and thereby eliminate the problem of false energy. Foldback in interferometry is an unrelated phenomenon resulting in analogous false peaks, and will be discussed in a later section.

Slits are used in dispersive spectrometry to control the resolution of the bands. Two conflicting factors strongly influence the selection of the slit width, and the slit width must be carefully chosen to minimize the adverse effects of both factors. If the slits are made very narrow to achieve high resolution, the relative intensity of the energy reaching the detector is reduced, and the signal to noise ratio (S/N) and overall sensitivity are reduced. If the slits are widened to allow more energy to reach the detector, resolution is degraded, and a band with a narrow absolute band width will appear as a band with a wide spectral slit width. Sometimes resolution can be degraded so much when using wide slit widths that a sharp band can disappear completely.

The detectors used in infrared spectroscopy are tailored to both the instrumental technique used and the range examined. In general, dispersive instruments are radiation-limited and require sensitive detectors, while interferometric instruments require detectors with a response rate fast enough to detect and transmit rapid energy changes to a recorder. The most common detectors are summarized in Table 4, along with some of the more salient features. Array detectors are still being developed for multiplex analysis in the NIR, but thus far the cost of these detectors has proven to be prohibitive to widespread use.

A single-beam dispersive spectrometer measures the energy of radiation reaching a detector after passing through a sample and a spectrometer for each spectral element along a scanning range. A dual-beam spectrometer chops the radiation from a source into two alternating pulses along two optical paths, one of which follows an optical path including a sample, the other of which follows an optical path as identical to the sample optical path as possible, but not including a sample. The second path is known as a reference beam. A recording device then measures the difference in energy registered by the detector. The spectra thus obtained can be presented

FALSE ENERGY

$$n\lambda = 2d \sin \theta$$

$n = 1, 2, 3 \ldots$

$d = 1000\, \ell/\text{CM}$

$\theta = 41.81°$

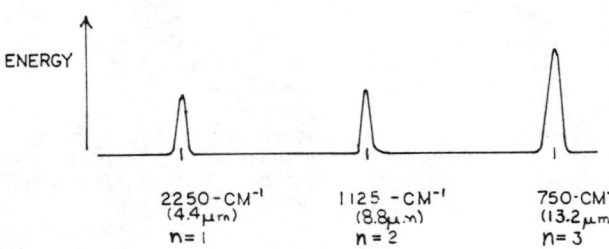

2250-CM^{-1}	1125-CM^{-1}	750-CM^{-1}
(4.4 μm)	(8.8 μm)	(13.2 μm)
n=1	n=2	n=3

WITHOUT FILTERING, A PURE AND SINGLE ABSORPTION AT 2250 CM^{-1} WOULD APPEAR IN THE SPECTRUM AS

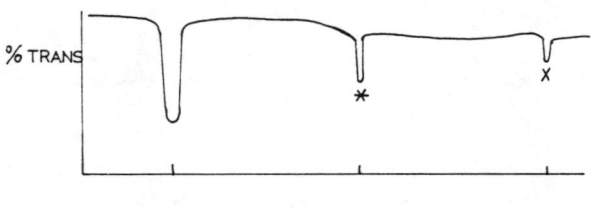

*FALSE ENERGY

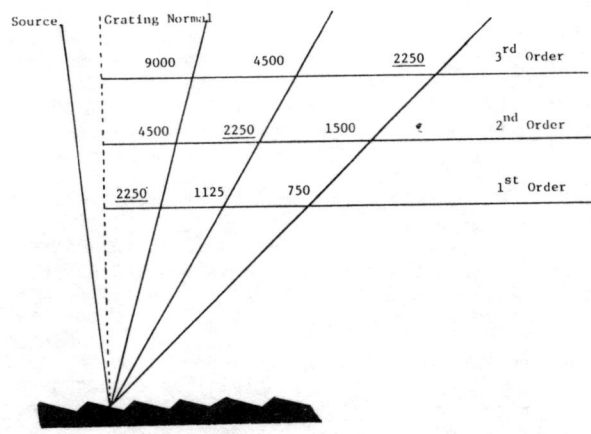

Fig. 3. False Energy Caused by Higher-Order Diffraction.
NB: Angles not Drawn to Scale for Demonstrative Purposes.

Table 4
DETECTORS PERFORMANCE

Detector	Operating Temperatures	Detection Range
TGS	Room Temperature	$5000 - \sim 30$ cm^{-1}
DTGS		
PbS	Liq N$_2$	Above 3000 cm^{-1}
PbSe		
InSb	Liq N$_2$	$20{,}000 - 1200$ cm^{-1}
Hg-Cd-Te(MCT)	Liq N$_2$	$3000 - 600$ cm^{-1} (or less)
Bolometer	Liq He (1.9°K)	$450 - 10$ cm^{-1}

Future Detectors Arrays
                  Par
                  IR-vidicon

as percent transmittance ($I_{sample}/I_{reference}$ * 100%) or absorbance (A = -log I_{sample}/I reference), each plotted against the wavenumber of the radiation. Thus one can more effectively compensate for fluctuations in the source, and source stability is not as crucial to reproducibility as it is for a single-beam instrument. The spectra also gives a much better indication of how much radiation a sample actually absorbs, as the energy profile of the detector and the source is factored out.

One problem with this arrangement which limits both the wavenumber and photometric accuracy of the bands is a result of continuous rotation of the grating during scanning. Because the grating is rotated continuously, the spectral element travelling along the sample beam is not exactly the same spectral element travelling along the reference beam, and the resulting comparison is not precise. Much of the inaccuracy could be eliminated by a stepwise grating movement, with measurements of the energies of the sample and reference beams made for identical spectral elements while the grating is stopped. One might hope to see instrumentation of this type in the future. When such instrumentation is available one can expect to see the wavelength precision advantage that interferometry has over dispersive instruments to be narrowed.

The two most common types of double-beam spectrometers are the optical null spectrometer, and the ratio-recording spectrometer. The arrangements of the two spectrometers are nearly identical, as can be seen by the schematic diagrams illustrated in Figure 4. The most important difference between these two spectrometers is the way in which the energies measured by the detectors are converted into transmission spectra.

The optical null spectrometer is the simpler and less expensive spectrometer. In the optical null spectrometer, an amplifier registers the difference in energy between the radiation in the sample beam and the reference beam, and drives a non-transparent comb or wedge into the reference beam until the energies are equal in both beams. The recorder, either a pen drawing on a chart or a computer, is attached to the optical comb, and as the optical comb is driven into the beam, the recorder registers less transmission. The main problem with this method is apparent when the sample

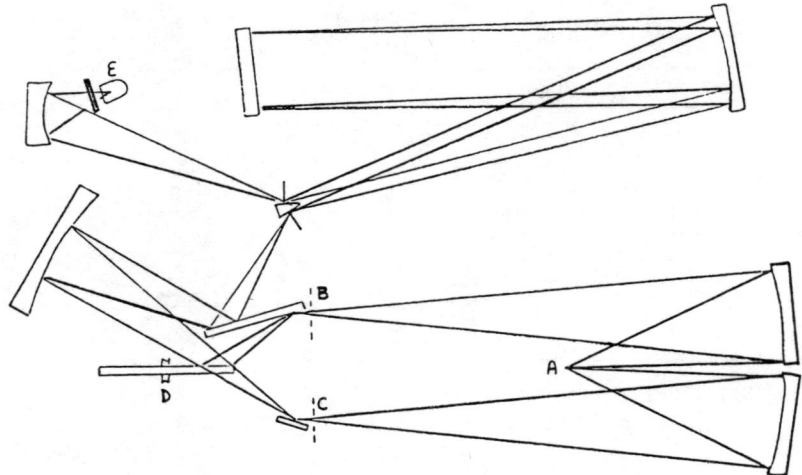

Fig. 4a. Optical Null Spectrometer.
A. Source; B. Reference Beam and Optical Wedge;
C. Sample Beam; D. Chopper; E. Detector

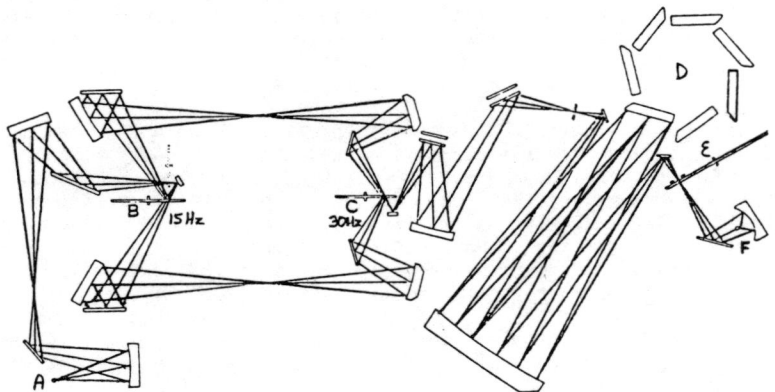

Fig. 4b. Ratio-Recording Spectrometer.
A. Source; B. Chopper; C. Chopper; D. Grating Rack;
E. Filters; F. Detector.

transmits very low levels of radiation. When this happens, very little energy travels through either beam, and it is very difficult to ascertain accurate transmission values.

The ratio-recording spectrometer avoids this problem by ratioing both sample and reference beam energy levels against a zero level when both beams are blocked. More photometrically accurate values are realized, but half of the energy is not used, and that can become an important factor in energy-limited experiments.

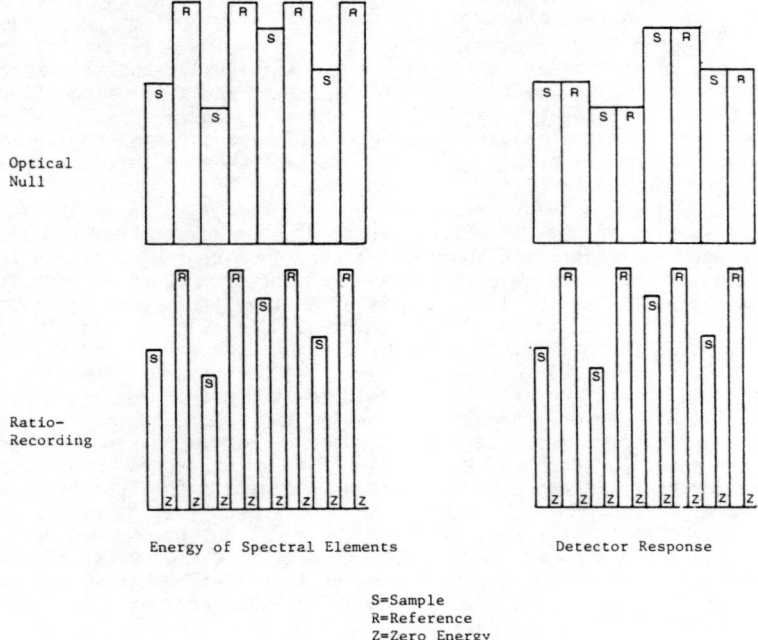

Fig. 5. Schematic Detector Response for Double-Beam Spectrometers.

A schematic plot of the detector response for both methods is presented in Figure 5.

As has been previously alluded to, Fourier Transform Infrared (FT-IR) spectroscopy approaches the problem of acquiring an infrared spectrum from a direction which is very different from the approach of dispersive spectroscopy. An interferogram does not resemble in any way a conventional spectrum; in fact, it requires a cognoscente to derive any data from an interferogram which might be useful in interpreting spectral anomalies. In order to present a conventional spectrum of percent transmission vs. wavenumber, the interferogram must be transformed digitally from a plot of detector response vs. optical path difference to transmission vs. wavenumber. In the early days of FT-IR, this calculation was laboriously assaulted by hand, and it was not until relatively recent times that fast Fourier transform algorithms, very fast computation times, and array processors have become available. It is therefore instructional to examine why early experimenters would take the trouble to develop such a strenuous technique.

A. A. Michelson produced the first interferometer nearly 100 years ago, and the interferometer which bears his name today is identical to the original in concept. Lacking a digital computer, he created an elaborately geared analog mechanism in an attempt to interpret the interferogram, although neither success nor failure is documented. His new instrumentation won only modest notice until workers began investigating the energy-starved FIR region of the spectrum. It will be remembered that the slits on a dispersive instrument severely limit the amount of radiation striking the detector, and it can be seen from Figure 2 that the sources do not emit much radiation in the FIR region to begin with. Since interferometry measures the energy of all frequencies emitted by the source simultaneously, and the entrance aperture is not limited by a slit, interferometry was a tempting alternative to investigators of the FIR region. Historically, the need for FT-IR has been the inverse of the amount of energy available: the higher wavenumber regions of mid and NIR have more energy available, hence there has been less incentive to develop FT-IR for these regions.

Two other factors influenced the slow application of FT-IR to the acquisition of spectra in the mid and NIR regions. The first hindrance arose from the need to perform the Fourier Transform, which requires time-consuming calculations. For a set resolution, the FIR spectrum requires much fewer calculations, and thus the transform is a relatively easier task. For instance, at 1 cm^{-1} resolution, the FIR, from 200 to 10 cm^{-1}, has 190 spectral elements, while the mid-IR, from 4000 - 200 cm^{-1}, has 3800 spectral elements, and the NIR, from 20,000 to 4000 cm^{-1}, has 16,000 spectral elements. The task of transforming the data is 400-fold for the mid-IR region, and 6400-fold for the NIR region, compared to that of the FIR region. The second hindrance arises from mechanical intolerances. Sample points must be acquired frequently enough to avoid fold-back, which occurs when a cosine wave of frequency ν/n can be drawn through the data points sampled from the cosine wave of frequency ν. Higher wavenumbers mean that more data points must be collected to preclude fold-back. The collection of more data points in turn requires a much more precise instrument, and higher precision leads to mechanical intolerances and operating difficulties. What made it worthwhile for early experimenters to overcome the impediments of mechanical intolerances and difficult calculations?

P. G. Fellgett, an astronomer, was the first researcher to investigate interferometry as an infrared technique. He realized that because all the frequencies of radiation were detected simultaneously, scanning times, which were notoriously slow in the far IR region, could be made much faster. It was later pointed out by P. Jaquinot that because the resolution of the spectrum is not determined by any slit widths, an aperture can be used, and much more energy is made available to energy-limited experiments. The triglycine sulphate (TGS) detector, a fast, fairly linearly responsive detector which could operate at room temperature, made detection in the mid-IR much more feasible. Once an interferometer was designed for FIR, all that was required for development of a mid-IR and NIR instrument was more precise engineering and faster methods of computation. With the development of the Cooley-Tukey algorithm, and the explosion of micro-circuitry, expansion to mid-IR and NIR depended mostly on the adaptation and optimization of the individual components of the interferometer to those regions.

Interferometry uses some optical principles which do not have immediately apparent relevance to infrared spectroscopy, so these principles will be explained qualitatively. The most basic aspect of interferometry is, appropriately enough, simple interference of monochromatic radiation. Any two waves will interfere with each other if their optical paths intersect. Two sources of radiation with the same wavelength and amplitude lead to the simplest case of optical interference. If the radiation happens to be in phase, the amplitudes will interfere constructively, and the resultant amplitude will be two times as great. If the radiation is out of phase by $(2n+1/2)\lambda$, the amplitudes will interfere destructively and cancel each other out. At intermediate phase differences, the amplitude can be calculated by the equation $0.5 (1 \pm \cos 2\pi\Theta/\lambda)$, where Θ is the phase difference. A coherent radiation interferogram, a plot of amplitude vs. phase for the interference of two sources of identical radiation is presented in Fig. 6a.

INTRODUCTION TO DISPERSIVE AND INTERFEROMETRIC INFRARED SPECTROSCOPY   37

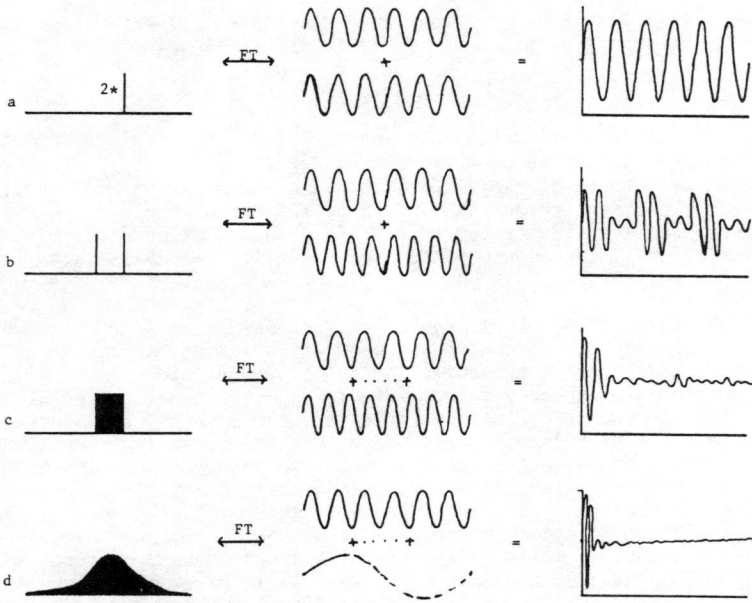

Fig. 6. Interferograms from
a. Coherent Radiation, e.g. Laser; b. Two Frequencies;
c. Narrow-Band, Constant Amplitude; d. Broad-Band,
Varying Amplitudes.

The same principle is used in dispersive spectroscopy to measure the thickness of an infrared sample cell or a film sample. As is illustrated in Fig. 7, some of the radiation striking the sample is transmitted without any reflection, while some of the radiation is internally reflected at the interfaces before passing through the sample. The transmitted radiation and the 2n-times-internally-reflected radiation interfere, and interference patterns appear in the spectra. The patterns are spectral artifacts, not vibrational or rotational bands, and are entirely a function of the thickness of the sample. They are very apparent in almost every polymer spectrum,

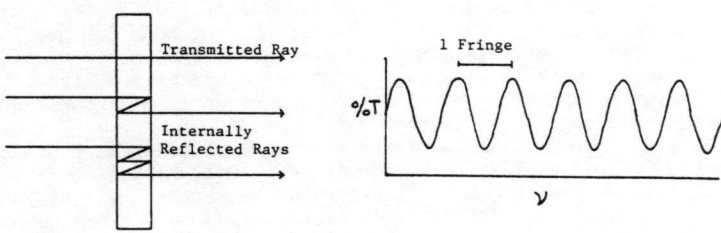

Fig. 7. Use of Internal Reflectance to Determine Cell or Sample Thickness.

but they can be inhibited by preparing wedge-shaped polymer films, thereby preventing internal reflection. Sample thickness is calculated by using the equation

Thickness=#complete fringes/$2\Delta\tilde{\nu}$.

It should be noted that the calculation of thickness by this method is linear with respect to wavenumber, not wavelength.

When two sources of radiation with different wavelengths interfere, the interference pattern is less simple. A beat pattern results from such interference, as is illustrated in Fig. 6b. The function describing the amplitude of the interference pattern of two different wavelengths is $0.5[\sum_{n=1,2}(1.0 + \cos 2\pi\Theta/\lambda_n)]$.

The pattern becomes less regular as more wavelengths are allowed to interfere. For instance, a continuous band of radiation results in an interferogram such as the one presented in Fig. 6c. Ultimately, when all the radiation from a black-body source interferes with itself or radiation from an identical source, the classical interferogram emerges, as illustrated by Fig. 6d. The Michelson interferometer was the first instrument designed to split broadband radiation into two separate paths, change the pathlength of one of the paths in order to induce an overall phase difference, and recombine the radiation of the two paths to cause interference. In this scheme, radiation is recombined before it passes through the sample and strikes the detector, while in a dispersive spectrometer, chopped radiation does not recombine. Even though both beams will share part of the same optical path, there is a time lag between detector responses to each beam.

A schematic diagram of a Michelson interferometer, which is still the most common interferometer, is illustrated in Fig. 8a. It consists of a fixed mirror, a movable mirror which can be displaced perpendicularly to the fixed mirror, and a beamsplitter set at 45° from the initial position of the movable mirror. The function of the beamsplitter is to transmit 50% of the radiation from the source to the fixed mirror, and 50% of the radiation from the source to the movable mirror. The composition of the beamsplitter depends on the spectral region under examination. Appropriate beamsplitters are tabulated in Table 5. For instance, in the mid IR region, a germanium/KBr beamsplitter is often used. The germanium reflects radiation, while the KBr transmits most of the desirable radiation. A compensator plate is necessary in some beamsplitter arrangements to ensure that both paths travel through the same optical materials before being recombined. It should be noted that in addition to all the absorbed and otherwise lost radiation, 50% of the radiation is lost when directed back to the source.

Phase differences are induced by careful linear movement of the movable mirror. As has been mentioned before, mechanical tolerance is very severe, and it is mostly owing to the constraints imposed by the need for precise motion of the movable mirror. In order to maximize precise movement of the movable mirror, an air bearing is usually employed. Lasers are used to make very precise measurements of mirror location, and often to make constant self-adjustments in mirror orientation.

As can be seen from Fig. 8a, a mirror displacement of x cm results in an optical path difference of 2x cm for instruments with Michelson interferometers. Instruments with Genzel interferometers, achieve an optical path difference of 4x. Resolution is inversely proportional to optical path differences, so it can be seen that the Genzel interferometer achieves the same resolution for only half the mirror travel of a Michelson interferometer, an important feature when mechanical tolerance is considered. The schematic diagram of the Genzel interferometer is depicted in Fig. 8b. The detector then measures the amount of energy at discrete intervals of mirror travel,

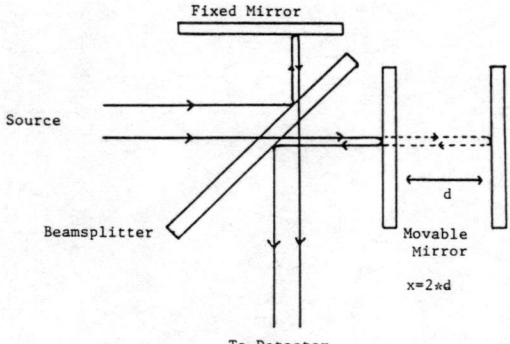

Fig. 8a. Schematic Diagram of a Michelson Interferometer.

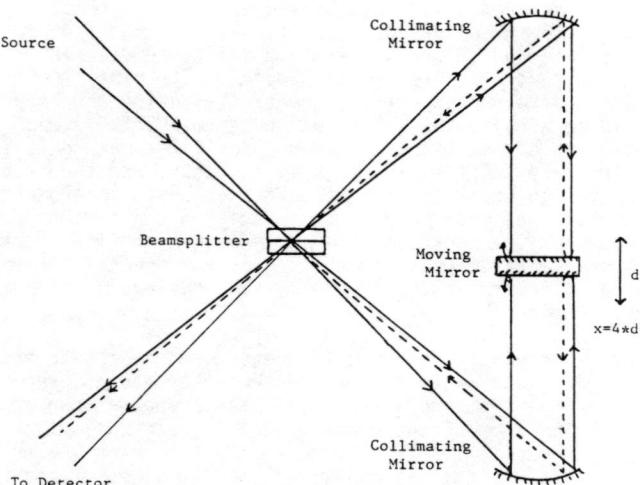

Fig. 8b. Schematic Diagram of a Genzel Interferometer.

and the interferogram is transformed into a conventional spectrum using the Fourier relationship

$$B(\tilde{\nu}) = 2\int_0^\infty I(\delta) e^{-2\pi\tilde{\nu}\delta} \, d\delta$$

usually with the aid of a dedicated microcomputer.

The spectrum generated from the transformed interferogram is a single-beam sample spectrum, and like the spectra obtained using a single-beam dispersive instrument, its general shape is determined by the energy profile of the source and the detector. A double-beam spectrum can be obtained by digitally ratioing the single-beam sample spectrum against a single-beam reference spectrum, which should be as identical as possible to the single-

Table 5

Beamsplitters

Region	Wavenumber Range	Materials
Near infrared	20,000 - 3,000	Quartz
	10,000 - 2,000	CaF_2
Mid-infrared	4,000 - 400	KBr
	800 - 200	CsI
Far infrared	650 - 100 down to 10 cm^{-1}	mylar, 6 etc.
	300 - 30 cm^{-1} down to 10 cm^{-1}	wire mesh etc.

beam sample spectrum, with the obvious exception of sample absorptions. Reference spectra can be obtained two ways, each of which has subtle advantages and disadvantages. The first method more closely resembles dispersive double-beam spectroscopy, and uses a mirror to switch radiation from a sample path to a reference path and back again after a certain number of interferograms have been collected. In this way, fluctuations in atmospheric absorption and source and general instrument drift are distributed most evenly between the sample and reference spectra. The other method is to acquire sample and reference spectra sequentially, using the sample path without a sample to acquire the reference spectrum. In this way, subtle path differences are not as great a factor. The first method is favorable to long scans in which drift is most likely to occur, and the second method is favorable to short scans taken after adequate instrument stabilization has occurred, and when very minor changes in the optical path would change the outcome of the spectrum. FT-IR is such a sensitive technique that sometimes the relative peak intensities of two spectra taken 2 seconds apart can change by several percentage points. Thus every attempt must be made to achieve reproducibility.

In addition to the infrared interferometer used to create the infrared interferogram, there is a white light interferometer and a laser interferometer. The white light interferometer tells the computer when to start data collection. The time of the commencement of data collection is more important than might be initially assumed. Interferograms are not perfectly symmetric, so data points must be collected on both sides of the centerburst in order to perform the necessary phase corrections. When interferograms are co-added, the data points of each interferogram must correspond as nearly as possible. As data collection does not begin with an easily recognizable event such as the infrared center-burst, a white light center-burst from an interferometer in series with the infrared interferometer triggers the signal to commence data collection. White light is used because the broad, short wavelength/high frequency radiation results in a very sharp, precise center-burst, with virtually no other spectral features which might erroneously trigger a signal to collect data.

The laser interferometer is used to measure the intervals between data point collection within each individual interferogram. The coherent monochromatic light yields an interferogram very unlike that produced by white light. The laser interferogram is virtually a cosine wave, with the period determined by the wavelength of the laser. The most commonly used laser is the helium-neon (He-Ne) laser, with a wavelength of 632.8 nm or 0.6328 μm.

The laser can generate a signal to measure an intensity datum every time the amplitude of the cosine wave is zero. The event is known as a zero crossing, and it occurs twice during the period of each cosine wave, or once every 0.3164 um for the He-Ne laser. A data point collected every 0.3164 um enables data acquisition up to 15,804 cm^{-1}, so when a routine infrared spectrum is being acquired to about 3800 cm^{-1}, data points need be taken only once every four zero crossings.

As has been mentioned before, there can be certain advantages to FT-IR spectroscopy. Although comparisons are odious, they are useful to the spectroscopist trying to decide if the inconvenience of interferometry is worth the theoretical improvement in detection limit, S/N, resolution, or precision desired for a particular experiment. Three main advantages to FT-IR are conventionally cited, although several other aspects of interferometry often lead to useful improvements in the quality of the spectra.

The most important, although often grossly overestimated advantage to FT-IR is the advantage of multiplexing, known as Fellgett's advantage. As all frequencies are measured simultaneously in interferometry, Fellgett's advantage can be used to either acquire spectra much faster, or to improve the S/N ratio for a given resolution. For instance, the 4000-400 cm^{-1} region has 3600 spectral elements of 1cm^{-1} resolution. Assuming that all other factors are equal, the interferometer can acquire the spectrum in 1/3600 the time required of a dispersive spectrometer. Alternatively, because the noise in IR detectors is essentially random, and increases as the square root of the number of scans N, and signal increases proportionately to the number of scans, 3600 scans can be co-added by the interferometer, and the S/N will be improved by $N/\sqrt{N} = \sqrt{N}$, or by a factor of 60. It can be seen that as the number of spectral resolution elements is decreased, either by sacrificing resolution or by investigating a smaller region of the spectrum, Fellgett's advantage is correspondingly reduced. Furthermore, not all parameters are equal for the two techniques, so Fellgett's advantage is often diminished still more.

The throughput, étendue, or Jacquinot's advantage is a measure of the relative amounts of energy allowed to reach the detector. It is a mistake to compare slit areas with apertures; the actual ratio is more closely approximated by comparing the area of the collecting mirrors in the interferometer with the area of the grating in the dispersive instrument, or comparing the f-numbers of the cones of radiation. Figure 9 shows cones for three instruments: the theoretical Jacquinot advantage enjoyed by the interferometer is $(11/2)^2=30$ times as much energy is available to the f11.0 instrument, and $(5.6/2)^2=7.8$ times as much energy as is available to the f5.6 instrument. As with Fellgett's advantage, Jacquinot's advantage rarely even approaches theoretical levels, owing to many factors. The reverse-Jacquinot advantage takes into account increased noise levels found in large-area detectors.

The Connes advantage makes use of the high precision of lasers to measure frequencies very precisely. There are no drawbacks to laser precision except for alignment problems, and most Fourier Transform software also includes calibration routines.

More detailed explanations of the advantages, and other advantages not mentioned here, along with the factors which reduce their benefits can be found in the references mentioned at the end of this chapter, and following chapters.

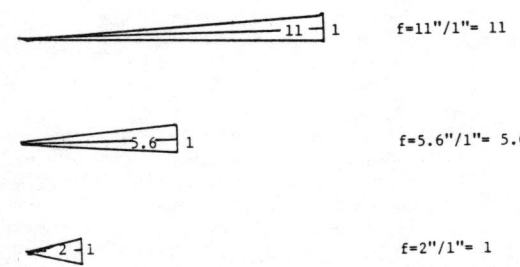

$$f = \frac{\text{focal length}}{\text{diameter}}$$

Increase in radiation grasp (etendue) or light throughput.

Interferometer vs. One type of Dispersion Instrument = $(\frac{11}{2})^2$ = 30 times greater

Interferometer vs. a Second type of Dispersion Instrument = $(\frac{5.6}{2})^2$ = 7.8 times greater

Resolution in dispersive instruments is dependent upon cone size - this is not the limiting feature in the interferometer.

Fig. 9. Relative Cones of Radiation.

Selected Bibliography

Bellamy, L. J. The Infrared Spectra of Complex Molecules, Vol. 1, 3rd ed. Metheun, Inc. New York, 1975.

Bellamy, L. J. The Infrared Spectra of Complex Molecules Vol. 2, 2nd ed. Metheun, Inc. New York, 1980.

These two volumes are the most useful treatises yet written on the subject of functional group identification using infrared spectra.

Bell, R. J. Introductory Fourier Transform Spectroscopy, Academic Press, New York, 1972.

Griffiths, P. R. Chemical Infrared Fourier Transform Spectroscopy, John Wiley & Sons, New York, 1975.

Two fine texts which cover basic theory as well as practical applications.

Smith, A. L. Applied Infrared Spectroscopy, John Wiley & Sons, New York, 1979.

Covers almost all aspects of infrared spectroscopy, yet has enough pertinent details to make it an extremely useful reference.

Hannah, R. W. and Swinehart, J. S. Experiments in Techniques of Infrared Spectroscopy, The Perkin-Elmer Corporation, Norwalk, 1974.

Manual for the beginning spectroscopist.

ADVANCES IN INSTRUMENTATION

Henry BUIJS

BOMEM Inc. Quebec City, Quebec, Canada

When one consults a text book on Fourier transform spectroscopy in order to understand what the technique consists of, one invariably becomes exposed to intricate optical and mathematical relationships. I would like to answer the question "what is a Fourier Transform Spectrometer and what is Fourier transform spectroscopy by relating F.T. spectrometers and F.T. spectroscopy to more traditional instrumentation and techniques?"

A brief historical overview

Spectroscopy was very fruitfully exploited when it was discovered that the light emitted by electric discharges, either in air or in glass tubes having very low vapour pressure, produces series of monochromatic spectral lines. From the positions of these lines a tremendous wealth of information regarding the structure of atoms was deduced.

This work, which commenced at the beginning of the 20th century, made use of several technological developments which occurred earlier. These are particularly: 1) the light dispersing properties of glass prisms-and later diffraction gratings-and, 2) photographic plates on which the dispersed spectrum can be recorded and subsequently measured.

Atomic spectroscopy using visible and U.V. radiation is still actively pursued today both for the purpose of studying outstanding problems related to atomic structure and identifying the presence of specific species.

The spectroscopic analysis and study of molecules is related to atomic spectroscopy in that spectral line positions provide information about the molecular structure. However, the non destructive method of transmission spectroscopy is much more prevalent for molecular species allowing the determination of characteristic spectral information not only for gas phase but also liquid and solid phase substances. The majority of rotation vibration absorption bands of molecules occur in the infrared region of the spectrum.

Routine analysis work, apart from atmospheric contamination problems, has been carried out mostly with solid and liquid samples. This tradition of analysis with solid and liquid samples in which spectral lines are spread out and diffused due to interaction by neighbouring molecules has led to a tradition of spectral analysis consisting of determination of molecular species by approximate band position, band shapes and combinations of bands present. The precise line position technique of atomic spectroscopy is generally not applicable.

As a result the traditional spectrophotometer for routine analysis work has had relatively low resolution, very limited wavelength precision but relatively well calibrated intensity capability while operating in the infrared.

The principle of the dispersive prism or diffractive grating spectrometer can be readily extended to the infrared and even to the far infrared. However the photographic plate for the recording and measurement of the spectrum is not available here. Early devices for sensing infrared radiation consisted of sensitive thermometric devices called bolometers. Bolometers do not form images and hence a spectrum must be recorded by sequentially displacing the bolometer to measure the intensity of radiation at different wavelengths.

A serious complicating factor in infrared spectroscopy arises at wavelengths longer than about 3.5 μm where the room temperature surrounding starts to generate significant levels of infrared radiation.

Towards the latter part of the second world war a program was initiated to develop an infrared spectrophotometer which could record automatically, on graph paper, the transmission spectrum of any small sample. The sample transmittance was compared against an empty beam of nearly identical optical path so that effects of room temperature infrared radiation and changes in the instrument response would be cancelled out.

These early spectrophotometers have developed into an entire family of spectrophotometers which for many years have played and important role in analytical chemistry. They have provided excellent wavelength coverage particularly in the mid and near infrared and adequate spectral resolution and wavelength calibration in order to differentiate between tens of thousands of different substances by means of subtle differences in the spectral bands.

Parallel to this wide proliferation of commercial infrared spectrophotometers a wide range of commercial and home made spectrometers, ranging from the very small to very large, occupying at times a large basement room, have come to be in order to study ever more subtle aspects of molecular structure or to decompose ever more complex mixtures of molecules.

The early Fourier transform spectrometers, which were nearly all home made, came into being to measure spectral features which were clearly <u>too weak</u> to measure with a grating or prism spectrometer and bolometer combination. They were small instruments having modest resolution and wavelength coverage.

Fourier transform spectrometers possess however an inherently much higher optical efficiency - the equivalent of the slitwidth being much larger than for a comparable dispersive spectrometer - as well as a much greater observation time efficiency - all spectral elements are simultaneously observed and measured by the bolometer in a unique Fourier transform code.

The optical sensor in Fourier transform spectrometers is a Michelson interferometer which has traditionally been considered a very delicate device. Furthermore the output signal from the bolometer does not correspond to the desired spectrum but must be transformed by means of a cosine Fourier transform computation.

To summarize, twenty years ago it was well established that Fourier transform spectrometers possessed a great advantage in sensitivity over more conventional spectrometers, but that the practice of Fourier transform spectroscopy was severely restricted because of the need for powerful computers to generate the spectral result and because the optical devices were very sensitive to perturbations.

A major breakthrough in the area of Fourier transform computation occurred in 1965 with the introduction of the fast Fourier transform algorithm by Cooley and Tukey. This had an immediate impact on Fourier transform spectroscopy allowing for the first time the measurement of extensive spectral intervals at very high resolution.

The introduction of the Helium-Neon laser had a significant impact on the evolution of the optical part -the interferometer- of the Fourier transform spectrometer. It allowed for the direct calibration of the interferometer mirror displacement which resulted in a much more precise measurement of the interferogram signal and hence a much improved spectral recovery.

Since the mid sixties there has been a continual evolution in Fourier transform spectrometers which has recently culminated in a very radical expansion in the applications of Fourier transform spectroscopy. This evolution has arisen particularly from the truly remarkable evolution in microelectronic circuitry and computer technology.

In the early sixties Fourier transform spectroscopy was fully dependent on the access to very large computers. With the evolution towards ever more powerful and smaller computers, even the most powerful Fourier transform spectrometers can now be operated with small dedicated computers.

Interferometer design and implementation has evolved to the point where all manufacturers of F.T. spectrometers have built-in sufficient rigidity and sufficient isolation from external perturbations such that their instruments can now operate reliably in a laboratory environment over extended periods.

We at Bomem have taken this evolution a step further by providing a completely actively controlled interferometer so that external perturbations such as temperature changes and vibrations are immediately corrected for. In addition to this immunity to perturbations, the active control, which includes automatic alignment acquisition, obviates the need for manual tuning of the interferometer mirrors; a task which is often tedious. Finally the precision of alignment that can be maintained during operation far exceeds that of more conventional interferometers thus leading to extensions of the technique not possible otherwise.

To summarize the F.T. spectrometer can fully emulate a more traditional grating or prism spectrometer both with respect to the ruggedness of the spectrometer and the stand-alone nature provided by means of compact low cost computing facilities. What must be emphasized immediately however is the fact that the F.T. spectrometer still possesses this very large advantage of sensitivity provided by the inherent optical and observational efficiency: these are called the Jacquinot and Felgett advantages respectively.

In addition to this emulation of traditional prism and grating spectrometers and the sensitivity factor, the F.T. spectrometer in its many present day forms brings to F.T. spectroscopy the possibility of a number of totally unique concepts.

Sensitivity Utilization

The most obvious result of the sensitivity advantage of F.T. spectrometers as compared to more traditional spectromoters is to obtain higher signal-to-noise ratio and, as a result, hopefully higher quality spectra. This higher quality does not follow automatically, but instead it is achieved only when we simultaneously put strong demands on non noise related characteristics such as frequency calibration, intensity calibration and instrument response reproducibility.

For standard mid I.R. spectra at modest resolution of say 1 cm^{-1} a signal-to-noise ratio expectation of $10^4:1$ is not unrealistic. Narrow spectral bands recorded at such S/N can be localized on a frequency scale to well within 0.01 cm^{-1}. In order to use such precision of frequency measurement, a corresponding calibration accuracy is needed. With some care absolute calibration accuracy of 0.01 cm^{-1} can be achieved on most commercial instruments: this absolute accuracy however cannot be taken for granted! On the other hand the frequency reproducibility on a single instrument is often much better.

The implication is as follows: spectral features occuring in spectra generated with the same F.T. spectrometer can generally be compared on a frequency scale which remains constant to within 0.01 cm^{-1}. Spectra generated on different F.T. spectrometers can generally be compared on a frequency scale which remains constant only to within about 0.1 cm^{-1}.

In view of the much greater resolution capability offered by the Bomem DA3 series F.T. spectrometers, a much more refined reference laser system is employed leading to much higher absolute frequency accuracy.

A similar situation arises regarding the intensity measurement precision and accuracy of calibration. Again from the S/N point of view we can often measure spectral intensities to 1 part in 10^4. However the linearity of the response between 0 and 100% transmission is by no means known to this precision. There are many factors affecting the accuracy of intensity measurement. A fundamental problem relates to the geometrical optical properties of the sample: the most severe example of this is the transmission characterization of a lens. When the sample does not contribute to geometrical optical abberation such as a low pressure gas in an absorption cell the reference of which is the evacuated cell spectrum, reasonable intensity calibration can be expected. In this latter case lineariy in signal processing and system stability in the spectrometer will determine the accuracy of the intensity scale. Generally this accuracy does not reflect the signal-to-noise ratio limit.

Where the F.T. spectrometer is suberbly adapted is in the area of small changes in intensity in spectral features. These may be very weak absorptions in virtually transparent samples or absorption bands superimposed on very strongly absorbing spectral regions where all amplitudes are near zero intensity.

As can be seen from the above, transferring higher measurement sensitivity into higher signal-to-noise ratio spectra does not automatically lead to higher quality spectra. It is to be noted that in many instances this is not a requirement.

More sensitivity?

From a simplistic point of view it would appear that F.T. spectroscopy technology cannot fully translate increases in sensitivity into correspondingly higher quality spectra. Despite this there is considerable interest in employing the most sensitive detection modules available.

Thanks to a strong interest shown by the defense research community in infrared detection, development in this area has been as rapid as it has been in F.T. spectroscopy.

This advance in detector technology however has had the effect of increasing sensitivity of traditional spectrometer systems as well. It can be shown in fact that traditional spectrometers gain more rapidly in sensitivity than F.T. spectrometers because of the inherently different manner of detector employment. It has also been suggested that grating spectrometers can be more sensitive than F.T. spectrometers: this would occur when large arrays of ultra sensitive detectors are employed to emulate the sensitivity of photographic film recording. This does not mean that the F.T. spectrometer would then immediately become obsolete.

What to do with the sensitivity

Most advances in F.T. instrumentation are related to answering this question. In fact the future of F.T. spectroscopy lies in the success with which we can answer this question. Fortunately in the process of perfecting the emulation of the traditional spectrometer the F.T. spectrometer has acquired features that make it unusually adaptable to many measurement situations. It is this adaptability which must be well understood in order to fully exploit F.T. spectroscopy.

Far Infrared Spectroscopy

The first commercial application of F.T. spectroscopy was for far infrared spectroscopy. At the outset there were three sound reasons for this:

1) The uncritical nature of the longwavelengths permits easy implementation of an interferometer with simple mirror displacement transducers and sensors.

2) F.T. spectroscopic sensitivity offsets the lack of availability of intense far I.R. sources.

3) F.T. spectroscopic sensitivity offsets the lack of availability of sensitive detectors.

Far IR Detectors

Recent advances in far I.R. instrumentation have been mainly in the area of detector improvement. Detection in the far I.R. is restricted to bolometric detection as opposed to the availability of photo conductive detection at wavelengths shorter than about 30 μm. An early standard for far I.R. detectors was the Golay bolometer detector. The periodic heating and cooling of a chopped I.R. beam is translated into expansion contraction of a small pocket of gas which in turn deflects a membrane. The membrane deflection is measured electro-optically and converted to the corresponding signal.

Much higher sensitivity and much lower noise can be achieved by means of a highly temperature dependent resistive device operated at low temperature: the liquid Helium cooled Ge bolometer. In most instances present day bolometers have an intrinsic noise characteristic which is well below the intensity fluctuation limit imposed by the statistical fluctuation in the photon flux incident on the detector. This is the so-called "background limited" performance limit.

To briefly indicate how this background limited performance limit is determined the number of photons N striking a detector per second can be calculated from the total power on the detector divided by the energy of the average photon. The uncertainty in this number N is given by $\sqrt{N}$. The power fluctuation then becomes $\sqrt{N}$ times the energy of the photon.

$$N = P/ch\sigma \qquad ch = 1.99 \times 10^{-23} \text{ watt.sec.cm}$$

$$\Delta P = \sqrt{N} \cdot ch\sigma$$

When the noise signal observed at the output of a detector corresponds to observing a fluctuation P the detector is said to be background limited.

This noise can be reduced only by reducing the number of photons per second striking the detector. Current bolometers incorporate liquid Helium cooled field of view masks and liquid Helium cooled optical bandpass filters so that only useful radiation is permitted to strike the detector element. In this manner <u>reduced</u> background limited performance is achieved.

Whereas it is usual to require many hours of signal averaging to obtain a reasonable quality spectrum in the far I.R. using a Golay detector, this same spectrum can be obtained in minutes with a well prepared liquid Helium cooled bolometer.

For very high sensitivity however the optical bandwidth may be as narrow as 75 cm^{-1} and the temperature of the bolometer must be reduced from the 4.2°K boiling point of liquid Helium to 1.5°K or less which is done by reducing the pressure above the Helium or even by employing a separate He^3 closed cycle cooling stage in the He^4 liquid.

This advance in the sensitivity of far infrared bolometers by extension increases the sensitivity of the F.T. spectrometer in this spectral region. The added sensitivity again may be used to increase the quality of spectra or permit extension of applications to either more rapid measurement or higher resolution measurement.

Far IR Interferometers

There have been several important advances in the interferometer optics for far infrared F.T. spectrometers. In the far infrared there is no satisfactory solid material than can be used for beamsplitters. The beamsplitter material in common use is a stretched film of Mylar of appropriate thickness so that constructive interference between front and back surface provides adequate reflection. Out of necessity such beamsplitters are highly wevelength dependent and as a consequence an extensive wavelength region requires the use of several different beamsplitters which must be interchanged.

Following a suggestion by Martin and Puplett the Mylar beamsplitter may be replaced by a polarizing beamsplitter which, in the far infrared, consists simply of a metallic wire grid. In a polarizing beamsplitter, perpendicular polarized components of the incident radiation are alternately transmitted and reflected. Upon being reflected back by the interferometer mirrors, the transmitted polarization is again transmitted and the reflected component again reflected. This has the effect of returning all radiation back towards the source only. An alternate approach is to rotate the polarization axes at the interferometer mirrors, in which case all radiation proceeds to the output. Since the radiation in the two arms of the interferometer is polarized in a mutually perpendicular manner, interference between the two components occurs only when the two polarizations are mixed via a 45° inclined polarization analyser. In order to achieve interference coupling between the two polarization axes, the incident radiation must first be polarized at 45° as well. As a result of the need for input and output polarizers, the polarizing interferometer has a theoretical efficiency which is 1/2 that of the non polarizing interferometer. However with an appropriate choice of wire grids, very uniform response can be achieved over a wide spectral range in the far I.R.: the higher efficiency of far I.R. polarizing grids largely offsets the fundamentally lower efficiency of the design.

The balanced two input - two output M.P. interferometer

The Martin Puplett polarizing interferometer lends itself readily to the implementation of a balanced two input - two output beam interferometer where the difference between the two beams is only the axis of polarization. This highly balanced system allows a direct interferometric ratio determination between the two inputs at both output beams separately. In the far I.R. this feature is currently exploited by Mather et al. and separately bu Gush in an attempt to measure the residual cosmic background radiation in outer space. This is believed to follow approximately a 3°K blackbody radiation distribution. The balanced dual beam eliminates the radiation contributed by the spectrometer and the actual measurement consists of a direct comparison of the space field of view with a calibrated 3°K blackbody source in the second input beam.

Dighham et al. have explored the possibility of extending the balanced dual beam polarizing interferometer approach to shorter wavelength by utilizing fine metal wire grids deposited on infrared transparent substrates as polarizing beamsplitters. The motivation here is to measure monolayer surface films by means of the difference in surface absorptivity between the two axes of polarization of incident radiation. The difficulty of this measurement requirement arises from the smallness of the surface absorption effect. The balanced dual beam polarizing interferometer would generate as its signal only the difference between the absorption by the two polarizations.

An alternate approach to balanced double beampolarizing interferometer is to modulate the polarization at a very high frequency and demodulate the detector signal so that the difference interferogram is observed only. This technique has been exploited very successfully by Nafie using circularly polarized light and modulating the direction of rotation with electro-optical crystals.

High resolution spectroscopy

In continuing to explore the sensitivity advantage of F.T. spectroscopy one finds that both for traditional grating spectrometers and F.T. spectrometers, the amount of transmitted radiation must be reduced as the resolution is increased. However, the throughput restriction for F.T. spectrometers follows as the square root of the increase in resolution whereas for the grating spectrometer this follows linearly. Furthermore, for a given constant detector noise characteristic, the observation efficiency advantage increases as the square root of the number of resolved spectral elements observed.

From the above it is evident that F.T. spectroscopy should lend itself well to high resolution spectral measurement. Work in the mid and late sixties by Connes et al. in the development of highly specialized high resoltuion F.T. spectrometers had led to a brilliant confirmation of these concepts.

In the late seventies we at Bomem were able to incorporate in a standard commercial F.T. spectrometer the capability of measuring spectra at very high resolution as well as at low resolution. The essential requirements to provide a range of resolutions are twofold:

1) the mirror displacement of the interferometer dictates the resolution in the computed spectrum as follows:

$$\Delta\sigma_{res.} = a/2L \text{ cm}^{-1}$$

where a is a line shape factor which may vary from 0.6 to 1.0 and L is the mirror travel from a position of equal path length.

i.e. $\Delta\sigma_{res.} = 1 \text{ cm}^{-1}$ L = 0.5 cm for a = 1

and $\Delta\sigma_{res.} = 0.004 \text{ cm}^{-1}$ L = 125 cm

2) given a mirror travel L the divergence of radiation passing through the interferometer must be limited. This is done by placing an iris diaphragm in the focal plane of the collimator analogous to the entrance slit in a grating spectrometer. The diameter of the iris is as follows:

$$d \leq \sqrt{2} \ f \ \sqrt{\frac{1}{\sigma L}}$$

where f is the focal length of the collimator, σ the frequency of interest and L the mirror travel. The diameter is constant for a given resolving power (R = $\sigma/\Delta\sigma_{res.}$).

$$d \leq \sqrt{2} \ f\sqrt{\frac{2}{aR}} \qquad 1/L = \frac{2\Delta\sigma_{res.}}{a}$$

In the earlier specialized high resolution F.T. spectrometers the interferometer design followed the principle of self compensation so that the interference fringes were immune to small errors in orientation of the moving mirror assembly. This leads to the use of either corner cube mirrors or so-called cat's eye retroreflectors in the arms of the interferometer. Both types of retroreflectors add complexity to the interferometer. Also the initial set-up of the interferometer must be done with great care. Finally the optical path through the interferometer becomes considerably longer than for a simple flat mirror interferometer. This has the net result that much radiation is lost due to vignetting when operated in the low resolution wide divergence mode.

In the Bomem F.T. spectrometer the simplicity of the flat mirror interferometer is maintained while at the same time compensating for errors in the orientation of the moving mirror by means of an active servo controlled tilt adjustment system. In this manner the spectrometer emulates an efficient small F.T. spectrometer for low resolution while at the same time providing a resolution capability which is higher than any grating spectrometer ever constructed. Hence we have been able to emulate in one F.T. spectrometer not just one class of grating or prism spectrometers but instead virtually all grating spectrometers ever constructed.

The observational efficiency advantage of F.T. spectroscopy presumes that detector noise is constant regardless of whether the detector is employed in a grating or F.T. spectrometer. This is true only for low quality detectors. Most high quality detectors, from bolometers in the far I.R. to photoconductive semiconductor detectors in the mid I.R., near I.R., and visible region, can achieve the "background limited mode", on F.T. spectrometers. In that case it becomes crucial to determine whether the radiation intercepted by the detector is radiation containing spectral information of interest or not. If the majority of the radiation carries no spectral information of interest the sensitivity advantage is greatly diluted. If the majority of the radiation consists of radiation containing spectral information of interest, the sensitivity advantage is maintained.

We at Bomem have extended the operation of the F.T. spectrometer from the very far infrared all the way to the visible and U.V. (5 cm^{-1} to 45 000 cm^{-1}). In the visible detector and U.V. the question of whether radiation received by the detector is significant or not becomes pertinent when we attempt to do Raman or ICP atomic emission spectroscopy. It is evident that some prefiltering is necessary however the extent of this filtering has not yet been determined.

Conclusion

The recent advances in F.T. spectrometer technology has seen tremendous extensions in 1) sensitivity in spectral intensity measurement, 2) resolution range, 3) total wavelength coverage and 4) wavelength or frequency calibration precision all accompanied by an ever greater ease and speed of operation.

These advances have led to very significant new applications in spectroscopy. It is my belief however that we are only at the beginning with regards to new applications. We have seen the full emulation of traditional analytical IR spectrophotometers by low cost FTIR systems. Because of the many new applications however we will probably see a proliferation of much higher performance FTIR systems at only moderately higher price.

METAL-MOLECULE INTERACTIONS WITH CHEMICAL AND BIOLOGICAL APPLICATIONS

INDUSTRIAL APPLICATIONS OF FOURIER
TRANSFORM INFRARED SPECTROSCOPY .

Jeanette G. GRASSELLI

The Standard Oil Company (Ohio)
Research & Development Division
Cleveland, OH 44128

Introduction

    Infrared spectroscopy is a widely used industrial tool for the structural and compositional analysis of organic, inorganic, or polymeric samples and for quality control of raw materials and commercial products. It is a relatively simple technique, non-destructive, versatile enough to handle solids, liquids and gases with a minimum of sample preparation, and accurate enough for both the qualitative identification of the structure of unknown materials and the quantitative measurement of the components in a complex mixture. An extensive body of literature on group frequency correlations exists as well as excellent reference spectral collections[A17]. Instrumentation has been reliable and low cost.

    But in spite of all these benefits, infrared spectroscopy has certain drawbacks which become more critical as the difficulty of the analytical problem increases. These drawbacks stem from the fact that infrared is an energy-limited technique. The energy distribution of the blackbody radiation of the IR source reaches a peak in the low wavelength region of the spectrum (2-5 µm), and falls off sharply and drastically at longer wavelengths. For routine operation there is generally more than sufficient spectral energy to obtain IR spectra useful for qualitative structural characterization and for the development of quantitative methods. However, situations frequently occur where there is not enough energy to accurately measure very weak or very strong bands necessary for an analysis. The bands could be weak because they are due to low concentrations of the component(s) of interest in an absorbing matrix, such as additives or impurities, or to extremely small amounts of sample, such as trapped chromatographic fractions. Or they could be bands which are naturally very weak but which must be optimized for some analysis. In addition, many IR spectrometers are not able to reliably record spectra of very thick materials.

    But the applications of infrared spectroscopy today have experienced an explosive "transformation" with the introduction of interferometric methods of obtaining infrared spectra and the subsequent mathematical processing of the interferogram via vast Fourier transform algorithms to recover the frequency spectrum. The "transformation" has given us impressive time and signal-to-noise advantages, as well as a whole new generation of instrumentation. A description of interferometry, as compared to dispersive IR instruments, is presented in the chapter by Grim and Fateley. In this chapter I will describe some typical industrial applications which demonstrate some of the special features of Fourier transform infrared spectroscopy. The chapter will follow this general outline:

Contents
- A. Introduction
- B. Data processing in FT-IR
- C. Applications to polymers
- D. Applications to coal, shale, minerals & heavy hydrocarbons
- E. Combined Techniques
    1. GC/FT-IR
    2. LC/FT-IR
    3. Evolved gas analysis/FT-IR
- F. Special techniques
    1. ATR, specular and diffuse reflectance
    2. Photoacoustic
    3. Matrix isolation
- G. Quantitative and trace analysis
- H. Inorganic
- I. Biomedical and pharmaceutical
- J. Interesting novel applications

An extensive bibliography from which many of the examples are drawn is included. The interested reader should refer to some of the specialized applications in the bibliography which are not discussed in this chapter. In addition, no reference will be made at all to the very interesting and active area of compound identification by automatic spectral searching, or structure matching. There just isn't room for everything in this kind of chapter!

Table I summarizes the instrument variables and sampling parameters that must be considered and reported when obtaining any infrared spectrum from a computerized spectrometer $^{(A6)}$.

TABLE I

Recommendations on Information to be Included with Published Spectra from Computerized IR Instruments

A. In Experimental Section
  1. Instrument Information

     Spectrophotometer manufacturer and model no.
     Any significant spectrophotometer modifications
     Detector information:
       Manufacturer and type
       Cut-off wavenumber (especially for Hg:Cd:Te)
       Size of detector element
     Resolution and apodization:
       Grating: slit width at ~1000 and 3000 cm^{-1}
       FT-IR: nominal resolution (1/optical retardation)
              apodization function
              number of data points per interferogram
              number of data points before the center burst
              single or double sided interferogram
              type of beam splitter
     Source type
     Instrument purged? evacuated? or neither?
     Electrical filters (high pass and low pass)

Table I (continued)

2. Scanning Measurements

   Total measurement time for data acquisition
   If signal-averaging has been performed:
     time per active scan
     time between starts of successive scans
     number of scans of sample (and reference, when necessary)
   Grating: starting and final wavenumbers and rate of scan
   FT-IR: mirror velocity (millimeters of retardation per sec)

3. Sampling Data for Digital Spectra

   Grating: wavenumber interval between sampling points
   FT-IR: retardation between sampling points, e.g., sampling frequency (kHz)
   For all instruments: dynamic range of ADC and word length of computer

4. Spectral Manipulation

   Smoothing:
     FT-IR: none, except by apodization of the interferogram
     Grating: state algorithm and extent of smoothing
   Interpolation:
     State algorithm used and number of interpolated points per independent
       data point
   Background correction:
     State type of correction used (e.g., single linear, multiple linear,
       parabolic, etc.)
   Derivative calculations:
     State algorithm used
   Fourier self-deconvolution:
     State parameters used
   Water vapor subtraction:
     State if done
   Fringe removal:
     State if done

5. Sampling Accessories:

   All accessories: give name and manufacturer
   Solution cells: sealed or demountable
                    pathlength (nominal or measured)
                    window material
                    solvent(s)
   Beam condenser: state linear condensation
   ATR: material used to construct IRE
                    angle of incidence
                    number of reflections (approx.)
   Specular reflectance (or R-A): angle of incidence
                                  reference material
   Diffuse reflectance: geometry of optics
                        reference material
                        neat sample or diluted with reference
   Diamond cell: diamond type
                    pressure (approx.)
                    pathlength (approx.)
                    diameter of open aperture
   GC-IR: length and diameter of light-pipe
          window material
          temperature of light-pipe
          temperature of transfer line
          column used

Table I (continued)

Photoacoustic detection:	nature of reference (source of carbon black)
	cup depth
	scan speed
	gas medium
	velocity drive
Long path gas cells:	number of reflections
	path length
	matched cells or single cell?
	nature of reference
Matrix isolation:	matrix gas
	sample concentration by volume
	pulsed or spray
	spray-on time
	annealing procedure
	temperature of matrix

B. Presentation of Spectra
  1. Numerical scales on abscissa and ordinate axes
  2. Difference spectra:
      Include spectra of starting materials
      State scaling factor
      State if ratio of two transmittance (T) spectra or scaled subtraction between two absorbance (A) spectra
  3. Ordinate expansion
      State if zero suppression was applied directly or was T spectrum converted to A, scaled and reconverted to T
  4. Background correction used

Data Processing in FT-IR

Some of the special advantages of Fourier transform infrared spectroscopy are given in Table II. These include energy limited, time limited, or signal-to-noise limited situations. Examples of each of these will be presented in the following discussion. For the spectroscopist who in the past has dealt with an analog instrument, there are aspects of the data processing in Fourier transform infrared applications which must be appreciated in order to not only take full advantage of the technique, but also to understand the measurement sufficiently for proper data manipulation and meaning-

TABLE II

SPECIAL ADVANTAGES OF FOURIER TRANSFORM IR

ENERGY LIMITED SITUATIONS
  OPAQUE SAMPLES < 1% TRANSMISSION
  IR EMISSION STUDIES
  VERY HIGH RESOLUTION REQUIREMENTS

TIME LIMITED SITUATIONS
  KINETIC STUDIES
  UNSTABLE COMPOUNDS
  REACTION OR CATALYTIC INTERMEDIATES

SIGNAL/NOISE LIMITED SITUATIONS
  TRACE ANALYSIS
  INTERFERING ABSORPTION
  ATMOSPHERIC SAMPLING

ful results (A1-4, 7-16). All of the excellent texts on FT-IR spectroscopy and many papers have been devoted to a discussion of the effects on the frequency spectrum from the mathematical operations which are carried out on the interferogram (B1,2,5,6). It is, for example, important to recognize that not only the resolution, but the width of observed infrared bands depend on the truncation (length of time that the interferogram is sampled) and the apodization function which is used. Figure 1 shows the familiar spectrum of polystyrene obtained on an FT-IR spectrometer with different apodizations, all other conditions being the same. Note the improvement in

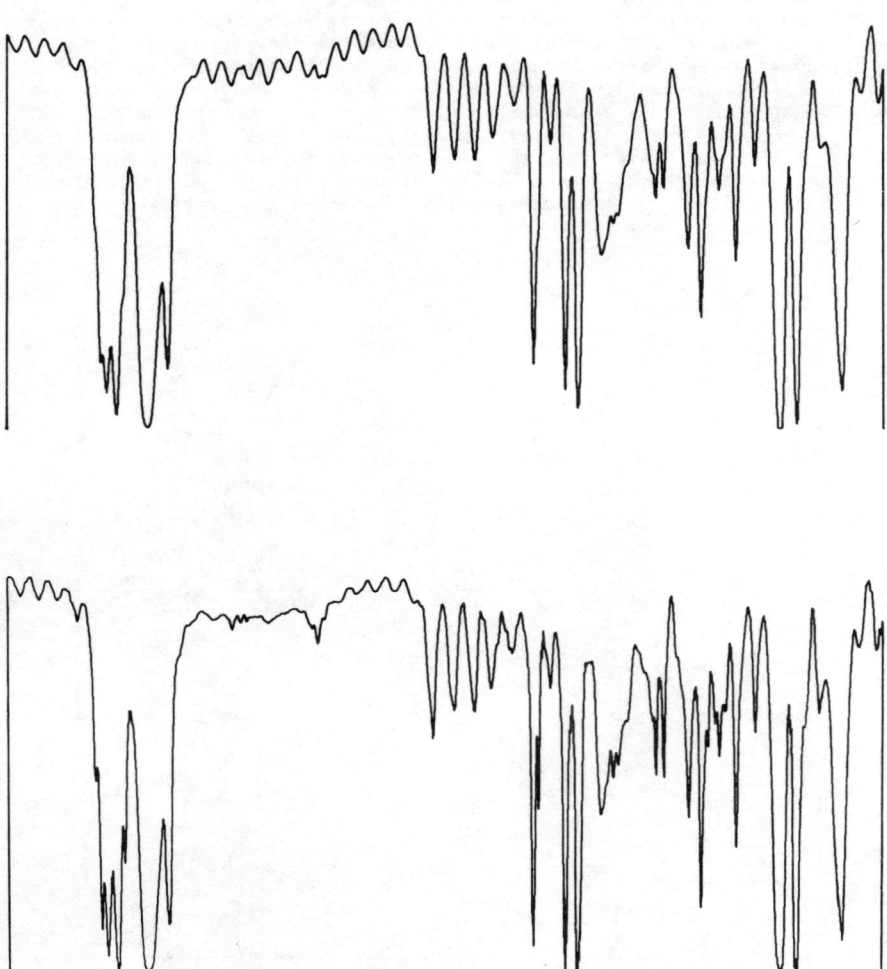

Figure 1   Polystyrene spectra measured using triangular (above) and boxcar (below) apodization. Each spectrum was measured using the same mirror travel, the only difference was in the position of point C of the apodization function.

resolution on several bands when boxcar truncation, rather than triangular apodization, has been used. Griffiths and Anderson (B1,B2) have pointed out the effect of apodization functions on absorbance subtraction data. Since subtraction is one of the most commonly employed types of data manipulation in FT-IR, it is important that the limitations of absorbance subtraction routines are well recognized. The measured spectra are the result of convolving the true bandshape of a peak with the instrument lineshape function of the spectrometer. Anderson and Griffiths thoroughly reviewed the effects on spectral subtraction experiments for both boxcar truncation and triangular apodization. They found that for relatively weak absorption bands (peak absorbance less than 0.7), the same resolution error on subtraction experiments was found if measurements were made at low resolution without apodization than if the interferogram were five times longer and triangularly apodized. However, when the peak absorbance is greater than one, if no apodization is applied the apparent peak transmittance can become negative and absorbance subtraction routines become impossible. If the interferograms are triangularly apodized, the apparent or measured peak absorbance will be smaller than the true peak absorbance, and is not proportional to it. Therefore, Beer's Law is not accurately obeyed and absorbance subtraction routines again become invalid. Anderson and Griffiths pointed

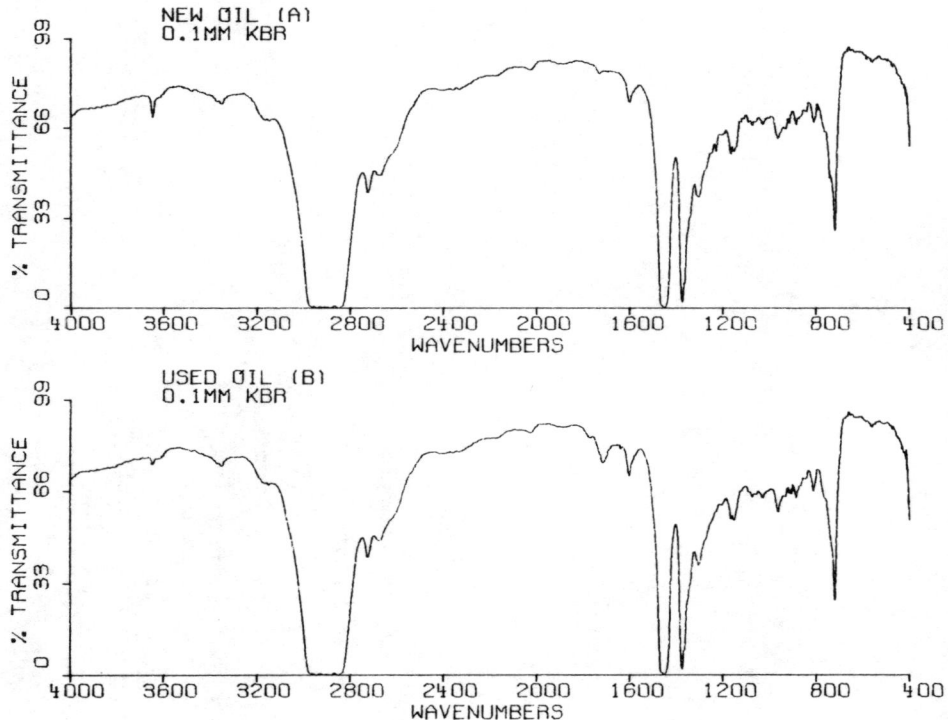

Figure 2   FT-IR Subtraction - Used Lube Oil Analysis
    (a)   New oil
    (b)   Used oil

out that different scaling factors would be required to minimize the residual features in difference spectra due to strong and weak bands, and that for any given scaling factor, some residual features will always remain in the difference spectrum.

For the practicing spectroscopist, however, absorbance subtraction becomes a powerful tool for the identification of minor components in strongly absorbing matrices. Figure 2 shows the analysis of a used compressor oil where contamination was suspected. The difference spectrum, Figure 2C, of the new oil, 2A, subtracted from the used oil, 2B, shows depletion (~60%) of the hindered phenol antioxidant additive (3650 cm^{-1}), appearance of carbonyl due to oxidation (1710 cm^{-1}), and olefin absorptions which are evidence of thermal cracking due to overheating of the oil.

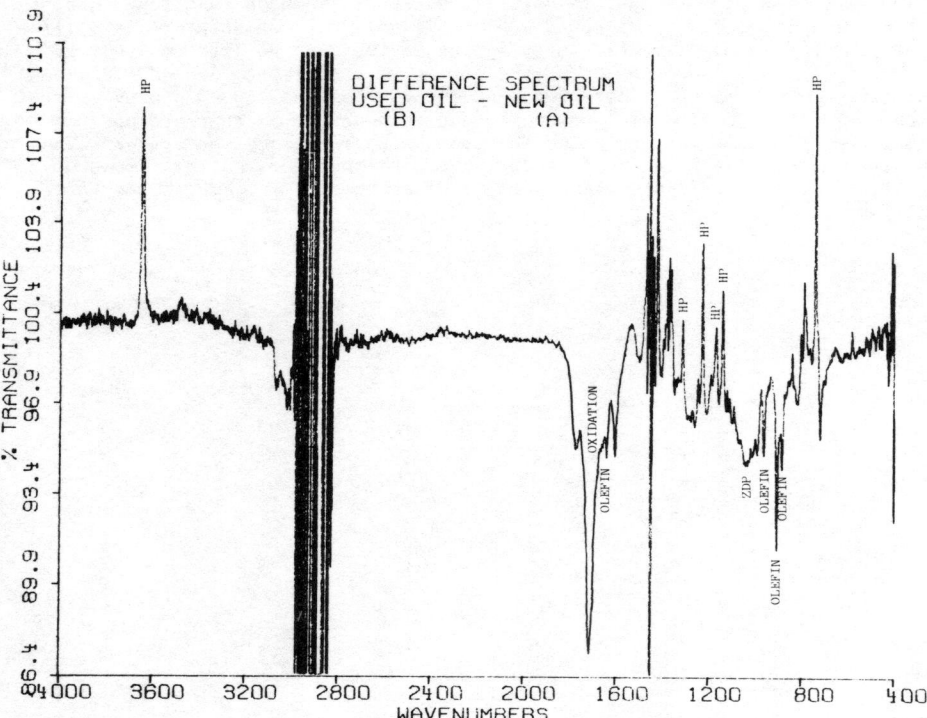

Fig. 2.   (c)   Difference spectrum   Used Oil minus New Oil

There are many examples of successful absorbance subtract in the literature, but it is instructive to also realize the limitations of the technique. To establish the limits of absorbance subtract on a complex sample typical to a petroleum laboratory, we did a series of controlled experiments to determine if we could detect small amounts of component additives in lubricating oils. Table III gives the blended lube oil composition and Figure 3A shows the starting infrared spectrum, obtained in a 0.1 mm cell. The first subtraction, Figure 3B, shows the lubricating oil minus the base oil. The difference spectrum plotted between 87 and 102%T shows the ester carbonyl at 1735 cm^{-1} due to the methacrylate viscosity index improver in the oil present at .08 wt.%; a 1700 cm^{-1} carbonyl band due to the alkyl succinimide; 980 and 660 cm^{-1} absorptions due to the zinc dialkyl dithiophosphate (ZDP); and bands noted on the spectrum due to the hindered phenol (2,6-ditertary butyl, p-cresol - DBPC). (This additive can easily be measured to the 0.25 wt.% level in normal lubricating oils using the 3650 cm^{-1} absorption of the "free" OH.) Although four out of five additives are detectable, it would be very difficult to specifically identify the kind of polyester used in the VI improver, or to be specific about the presence of a succinimide or of the diphenylamine in a total unknown spectrum. Three sequential spectral stripping subtractions were conducted: (1) to remove the ZDP by subtracting on the 980 cm^{-1} band; (2) to remove the hindered phenol by subtracting on the 1230 and 1155 cm^{-1} bands; (3) to remove the diphenylamine by subtracting on the 1295 cm^{-1} band. The final spectrum shown in Figure 3C shows good signal-to-noise, and certainly major functional groups of the remaining additives could be identified. But note that it is still not possible to specifically identify the VI improver at 1735 cm^{-1} nor to identify an alkyl succinimide as the source of the 1700 cm^{-1} band. Residual hindered phenol is still observed, and the ZDP band at 660 cm^{-1} plus several others in the 900-1100 cm^{-1} region have gone negative. We could not "balance" the spectral stripping factors, using reference spectra of <u>known</u> additives, to give a useful fingerprint region spectrum. Slight frequency shifts of some bands were also noted, indicating some intramolecular interactions.

TABLE III

BLENDED LUBRICATING OIL COMPOSITION

BLEND	WT. %
BASE OIL	99.0
HINDERED PHENOL	0.4
ZDP	0.11
POLYESTER	0.08
ALKYL SUBSTITUTED DIPHENYLAMINE	0.22
ALKYL SUCCINIMIDE	0.20

Applications to Polymers

FT-IR has found particularly wide application in the field of polymer analysis, not only because of the ability to look at intractable, thick, intensely absorbing materials, but also because of the ability to look at chemical and physical changes in the polymer structure as they are occurring. Koenig and coworkers (C3-7) have evaluated the interaction of antioxidants in butadiene rubbers (even those with carbon black filler) and the effects of various additives on cure. They have elucidated the crystalline and amorphous components of many polymer systems.

A particularly interesting example of surface chemistry of polymers was the examination of an E glass fiber coated with a coupling agent, vinyl trimethoxy silane (VTMS). Figure 4A shows the fiber and coupling agent, and spectrum 4B is the E glass fiber alone. The difference spectrum shown in 4C clearly identifies the VTMS coupling agent present on the surface.

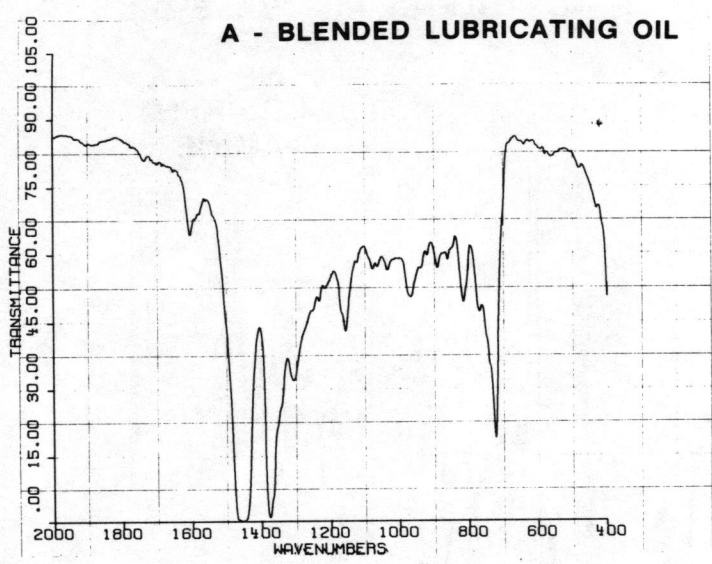

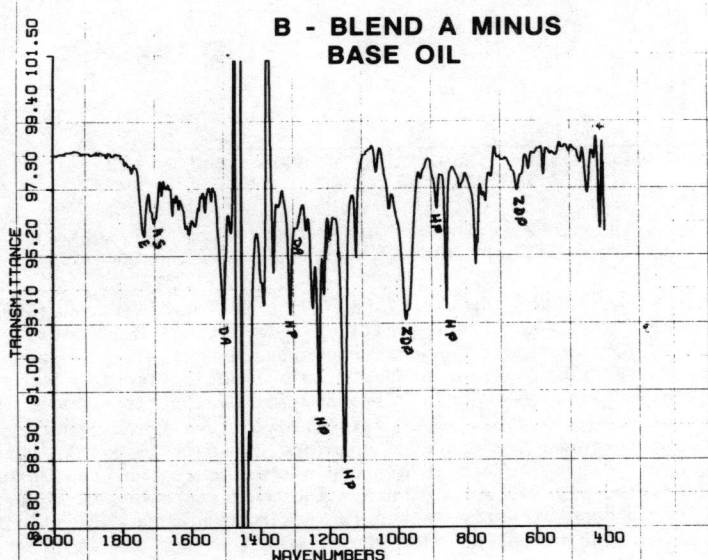

Figure 3 FT-IR Difference Spectrum
   (a) Blended lubricating oil
   (b) Blended lubricating oil minus base oil

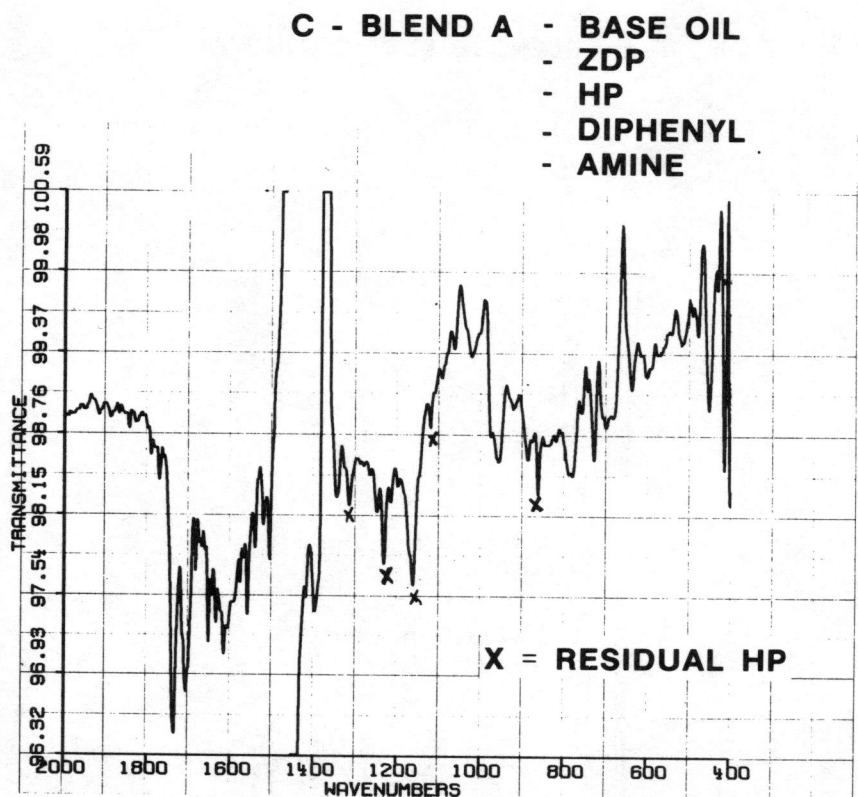

Fig. 3. (c) Blended lubricating oil after successive subtractions of base oil, and ZDP, HP, and diphenylamine additives

Several studies described the use of Fourier transform infrared spectroscopy to measure the extent of cure in fiber-reinforced epoxy composites (C2-9), and to follow time-dependent intensity changes of absorption bands characteristic of reactants or products in polymerization reactions (C8). For the epoxy resins, both thin films and internal reflectance spectroscopy techniques were used. Cure was followed in a temperature-controlled cell and infrared spectra were recorded at short time intervals throughout the cure cycle. The extent of cure was based on the epoxide ring absorbance at approximately 915 wavenumbers. The authors point out that the FT-IR method offers a nondestructive means for optimizing cure cycles in the laboratory and performing extent of cure measurements in a plant fabrication environment.

A typical industrial application of FT-IR in troubleshooting a polymer problem was described by Grasselli and Wolfram (A5). Orientation of an acrylonitrile/styrene copolymer film resulted in development of very small spots throughout the film. The spots appeared to be chunks of gelled resin, but it was also possible that they might be due to a contaminant, a small amount of homopolymer, or even a trapped liquid or gas. FT-IR subtraction spectra of a pinhole aperture section of the oriented film, subtracting the clear area from the spots gave a spectrum, Figure 5, that was easy to iden-

tify as polyvinyl acetate, an obvious contaminant in the polymer. Many
other examples to polymer analysis are described in the references cited in
Section C of the bibliography.

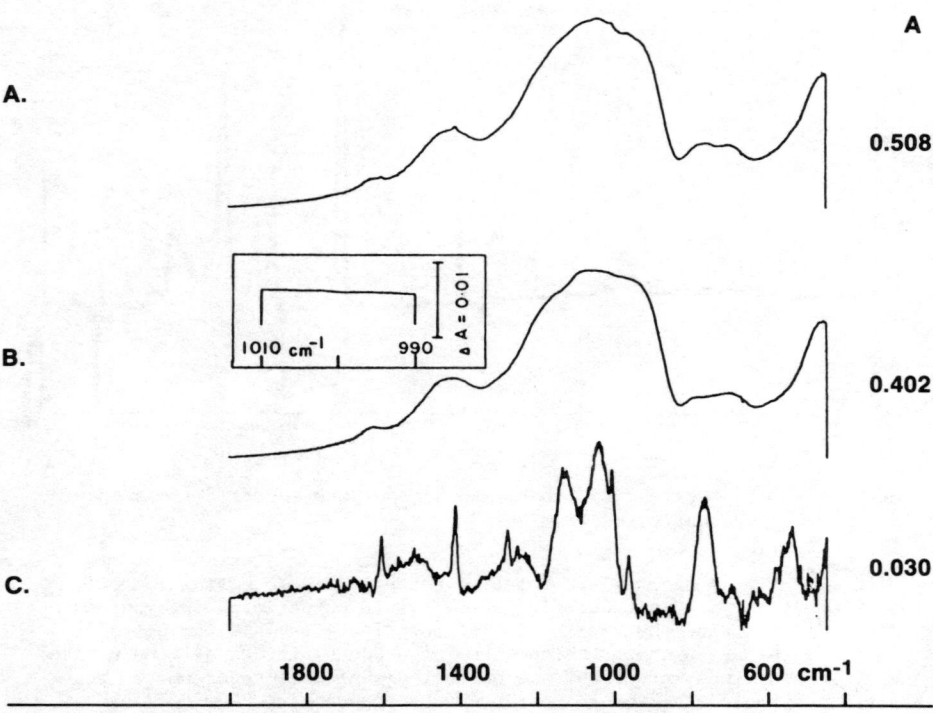

Figure 4   FT-IR Surface Analysis
    (a)   E-glass fiber/VTMS
    (b)   E-glass fiber
    (c)   Difference spectrum showing VTMS

Applications to Coal, Shale, Minerals and Heavy Hydrocarbons

Perhaps one of the most difficult samples for the industrial spectroscopist is black, hygroscopic, a compositionally complex structure including
huge organic and varied inorganic constituents - i.e., a description of
coal. There is no question that the interest in coal chemistry has inten-
sified recently due to the prospect that coal, which is a very abundant
natural resource, can be converted to liquid hydrocarbon fuels to feed our
energy intensive economy. The direct characterization of solid coal
structures by IR spectroscopy has been severely handicapped because coal is
opaque and is difficult to handle. A significant understanding of coal
structure has been provided by the work of Painter and Coleman at Penn State
(D1-8). They have looked at the mineral content of coal, as well as some of
the products of oxidation of coal, from which original structures can be
deduced.

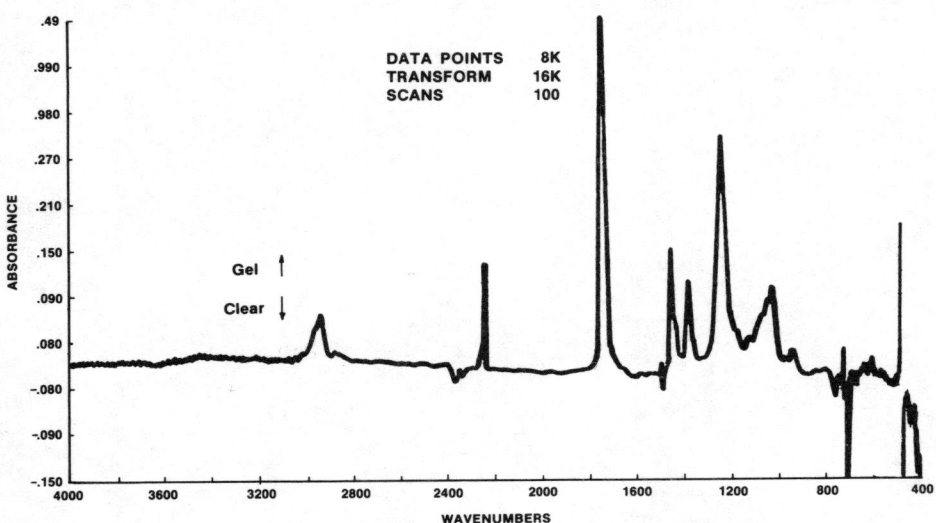

Figure 5 "Gelled" section of AN/S copolymer sheet; clear area subtracted from gel

In a similar manner, FT-IR has also been applied to evaluating oil shale properties[D9], particularly the oil yield potential. Solomon and Miknis[D10] describe the application of quantitative FT-IR for oil shale evaluation. Figure 6 shows the spectrum of a South African oil shale with mineral subtractions, after which a corrected shale oil spectrum can be observed. By measuring the aliphatic CH absorbance near 2900 cm^{-1} a correlation with oil yield, as obtained in a Fischer assay, was observed.

Combined Techniques

There is no question that one of the most exciting areas of application of FT-IR spectroscopy has come about because of the time advantage that is gained in the interferometric method for obtaining an IR spectrum, i.e. - the real time coupling of an infrared spectrometer to a gas chromatograph was finally realized. Not only the packed column GC separation, but now also capillary GC separations have benefited from the identification power of the FT-IR spectrometer. An excellent review of GC/FT-IR applications is by Erickson[E3], as well as in the special issue of the Journal of Chromatographic Science[E8]. For all these applications, special attention must be paid to the configuration and dimensions of the lightpipe and the transfer lines between the chromatograph and the IR. Several ways of manipulating and of presenting the infrared spectra have been developed. A particularly interesting example was that of a sample extracted from New Orleans drinking water (Figures 7A & B). Twenty microliters of a 10^6 concentrate of organics from the drinking water was injected into a 1/8" ID, 10' long glass column, 3% SP2100. A flame ionization detector was used, and it is clear that a very rich collection of organics is present. Peak 24 was selected to demonstrate the special power of FT-IR identification on the separated peaks. Spectra were taken starting at the leading edge of peak 24, and every two consecutive interferograms were coherently added while

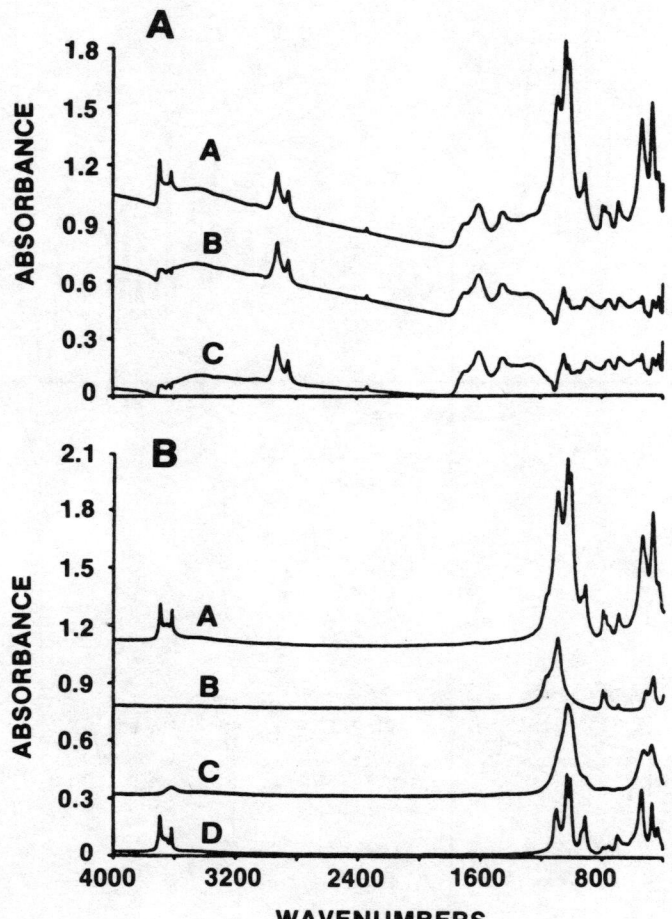

Figure 6　　Mineral subtractions and scattering corrections for infrared spectrum of South African oil shale. The abcissa is in units of absorbance for a sample density of 1 mg/1.33 cm^2.

(a)　A, whole shale; B, shale minus minerals; C, corrected shale

(b)　A, mineral spectrum, 49.0%; B, quartz, 11.0%; C, illite, 22.0%; D, kaolin, 16.0%

scanning through the peak. The resulting series of spectra are shown in Figure 7B. It is clear that at least three compounds are present in this peak and spectrum file number 1160-1 was especially different. However, with the use of gas phase reference spectra in the infrared and high resolution mass spectroscopy for confirmation, the compound was identified as a chlorodialkyl-N-(methoxymethyl) acetanilide. Several comparisons have been made of the advantages of GC/FT-IR and GC/MS (E[4,7,19]), or "hyphenated techniques" in general. It is clear that each of the combinations has special

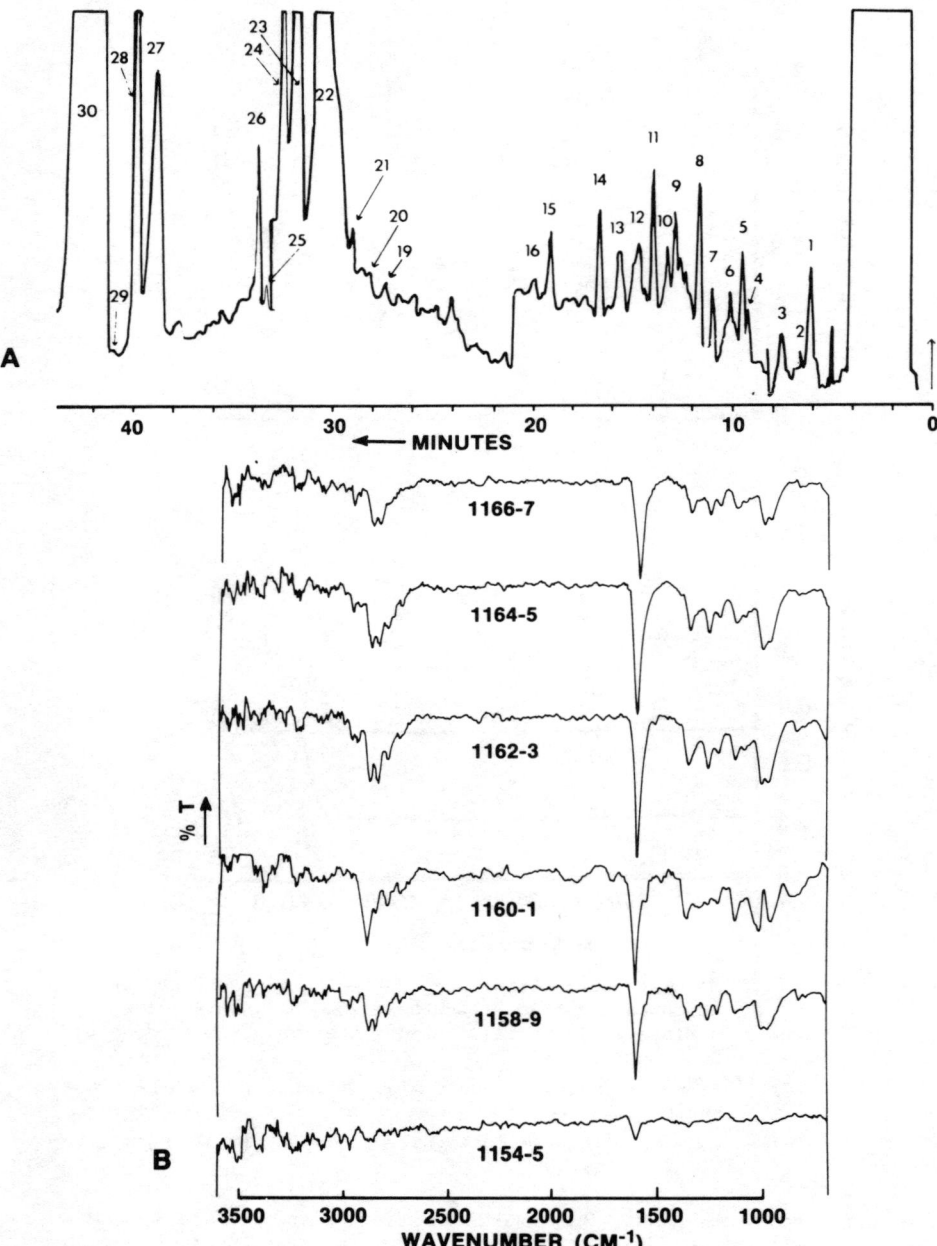

Figure 7 (a) Gas chromatogram with flame ionization detection of a 10^6 concentrate of organics from New Orleans drinking water.

(b) Spectra obtained from New Orleans drinking water sample, beginning on the leading edge of peak no. 24 (Figure 11) showing the rapid change in the spectra with time.

strengths, and that hyphenated techniques are now the rule rather than the exception.

Examples of capillary GC/FT-IR are now appearing rapidly (E5,6,9,10, 17,18). In Figure 8 the spectrum of approximately 20 nanograms of isobutyl methacrylate (which seems to be the calibrating substance for determining sensitivity in GC/FT-IR) is shown (E17). The spectrum was obtained from a 0.25 μl solution of isobutyl methacrylate in methylene chloride injected directly onto a capillary column. The column was 25-m glass, wide-bore, OV-101 WCOT. The lightpipe was an Accuspec model 26, 6 cm long, 2 mm in diameter, and with a volume of 0.19 ml.

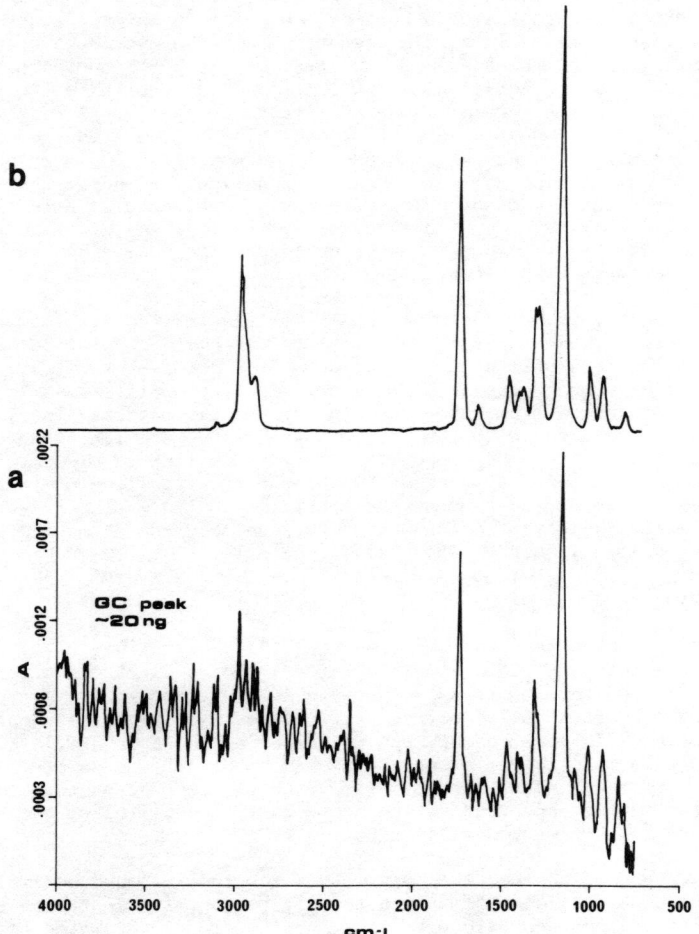

Figure 8 (a) Spectrum of approximately 20 ng isobutylmethacrylate
(b) Sadtler library spectrum automatically retrieved as closest match.

Liquid chromatography has not yet been interfaced in quite as successful a manner with FT-IR as has gas chromatography (E20). In general, the highly absorbing nature of most solvents used for LC has made the possibility of flow-through cells with direct examination of the LC fractions as they are eluting from the columns very difficult, even utilizing the capability of subtraction and optimization of detection limits. However, Griffiths and Kuehl (E11) have described a very effective off-line combination of HPLC with FT-IR, utilizing diffuse reflectance techniques. The eluent from the LC is evaporated onto KCL powder, and the infrared spectrum in diffuse reflectance can be obtained. Using a carousel arrangement of small cups filled with KLC, a semi, real-time apparatus has been designed.

It is often necessary to identify polymer structure of intractable crosslinked polymer samples or other insoluble filled materials, or to follow degradation mechanisms. In recent years techniques have been developed to observe the infrared spectra of gaseous effluents during polymer degradation. Liebman, Ahlstrom, and Griffiths (E15) published the degradation products from a programmed heat treatment of a polyvinyl chloride composite in air. Combustion products, including HCl, SO_2, CO_2, CO and CH_3COOH, were observed and their change in concentration could be monitored as a function of decomposition temperature and time. Lephardt and coworkers (E12-14) have elegantly coupled thermal gravimetric analysis with FT-IR and have pioneered applications of evolved gas analysis (EGA), particularly to combustion products of tobacco. The kinetics and mechanism of the combustion processes were elucidated.

The recent introduction of "portable" FT-IR instruments is making the combined thermal analysis/FT-IR interface an extremely popular one with great potential for looking at polymer products, catalyst samples, coal and oil rock. The rugged portable FT-IR's also open up applications in analyzing gaseous effluents from research or pilot plant reactors. The instrument can be mounted on a laboratory cart and taken to the chemical reactors throughout a facility. The FT requires only a 110 line to plug it in and about 10 minutes of warm-up. Figure 9 shows typical on-line monitoring of a reactor gas stream effluent using the mobile FT-IR (E2). The gaseous components are easily identified from this 20 scan spectrum at 4 cm^{-1} resolution. An Accuspec lightpipe was used as the cell.

Special Techniques

Griffiths and coworkers have pioneered the applications of diffuse reflectance spectrosopy using FT-IR (F2,3,8), and Powell (F14) has shown that DRIFT can be used to study surfaces. An excellent recent example that illustrates the potential of diffuse reflectance as a powerful sampling technique for FT-IR spectroscopy was described by Chase et al. (F1). They used DRIFT to obtain infrared spectra of paints directly on paper panels. The contribution of the binder was eliminated by spectral subtraction, and the products of the photodegradation were easily identified as the samples were exposed to accelerated weathering.

Another promising spectroscopic method for the infrared analysis of samples which do not lend themselves readily to normal preparation techniques is that of photoacoustic detection (F4,12,17,18). Examples of the application to catalyst, polymer, chromatographic sections or spots, and solid fuels such as coals, tar sands and shale oils, have been discussed (F5-7,9,10,13,16). One of the most novel applications of infrared photoacoustic spectroscopy is that described by Riseman et al. on conducting polymers (F15). Semi-conducting synthetic polymers, such as polyacetylene, have interesting physical properties, however are quite opaque and do not

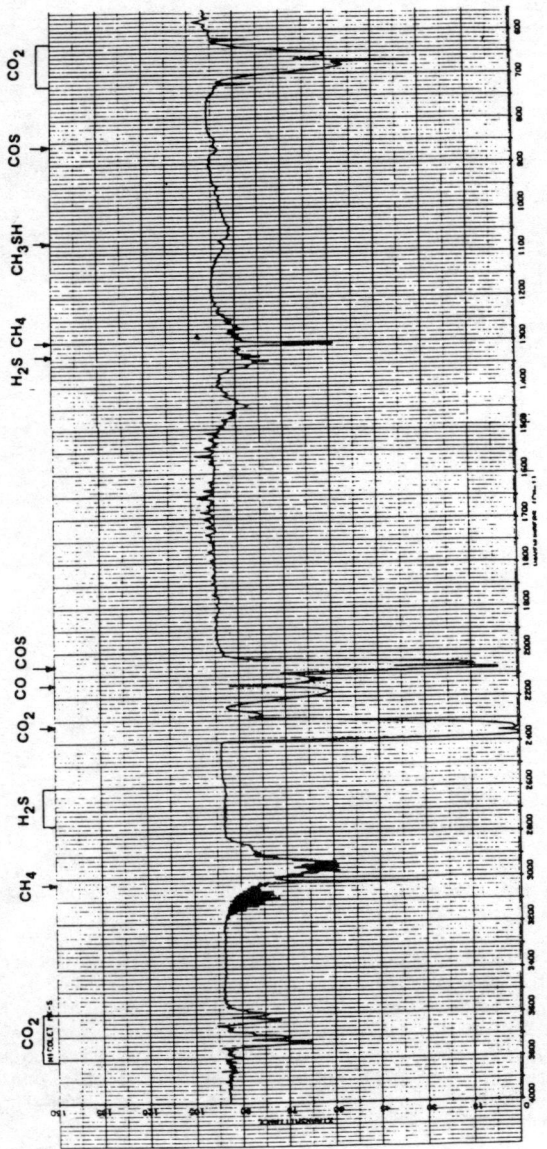

Figure 9   On-line monitoring of reactor gas stream effluent - mobile FT-IR

produce good reflectance spectra. Photoacoustic detection is a good alternative. Samples of undoped and n-doped polyacetylene were examined and spectral shifts and intensity differences were observed, which were interpreted in relation to the conducting properties.

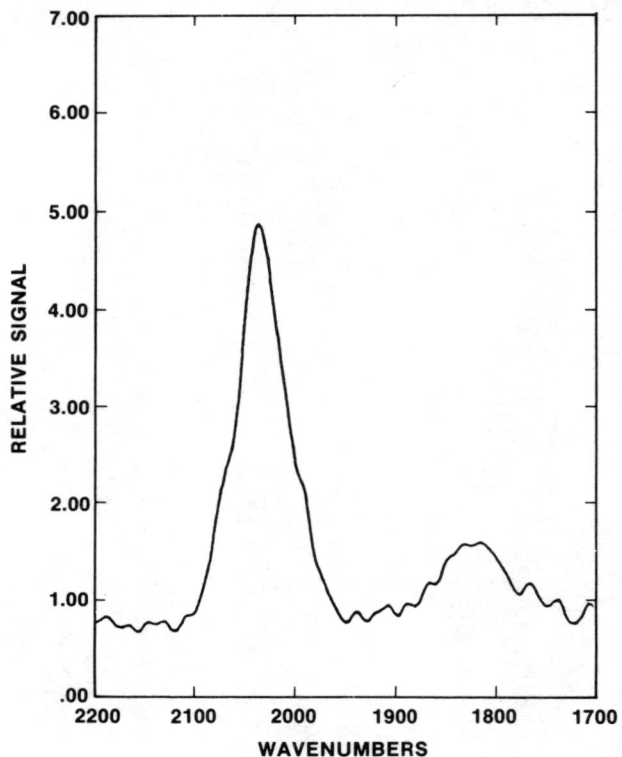

Figure 10   PAS/FT-IR, mid-infrared spectrum of CO on 5% Pt on Alumina Catalyst

Photoacoustic detection provides a convenient way of examining heterogeneous catalysts and observing molecules on the surfaces. Kinney and Staley [F6] have described a photoacoustic cell designed for temperature and atmosphere control in chemical studies. The capability to detect surface species is shown in Figure 10, the spectrum of CO on 5% platinum on alumina catalyst. High sensitivity is observed with relatively little sample preparation. Data collection times are short and the potential for observing in-situ reactions is certainly interesting.

Another special technique that deserves mentioning is the use of matrix isolation Fourier transform infrared spectroscopy, discussed recently by Mamantov et al. [F11,19]. These workers have demonstrated both the qualitative and quantitative applications of matrix isolation to the study of intermediates and as a way of obtaining highly resolved spectra for molecular structure determination. The coupling of this technique with chromatographic separation is also described. Particularly interesting applications to the detection of polynuclear aromatics is described and applied to an effluent sediment from a coking plant operation and to an aromatic fraction from a coal-derived crude oil.

Trace Analysis - Quantitative Analysis

I would be remiss to not emphasize in this overview of industrial applications of Fourier transform infrared spectroscopy the immense impact

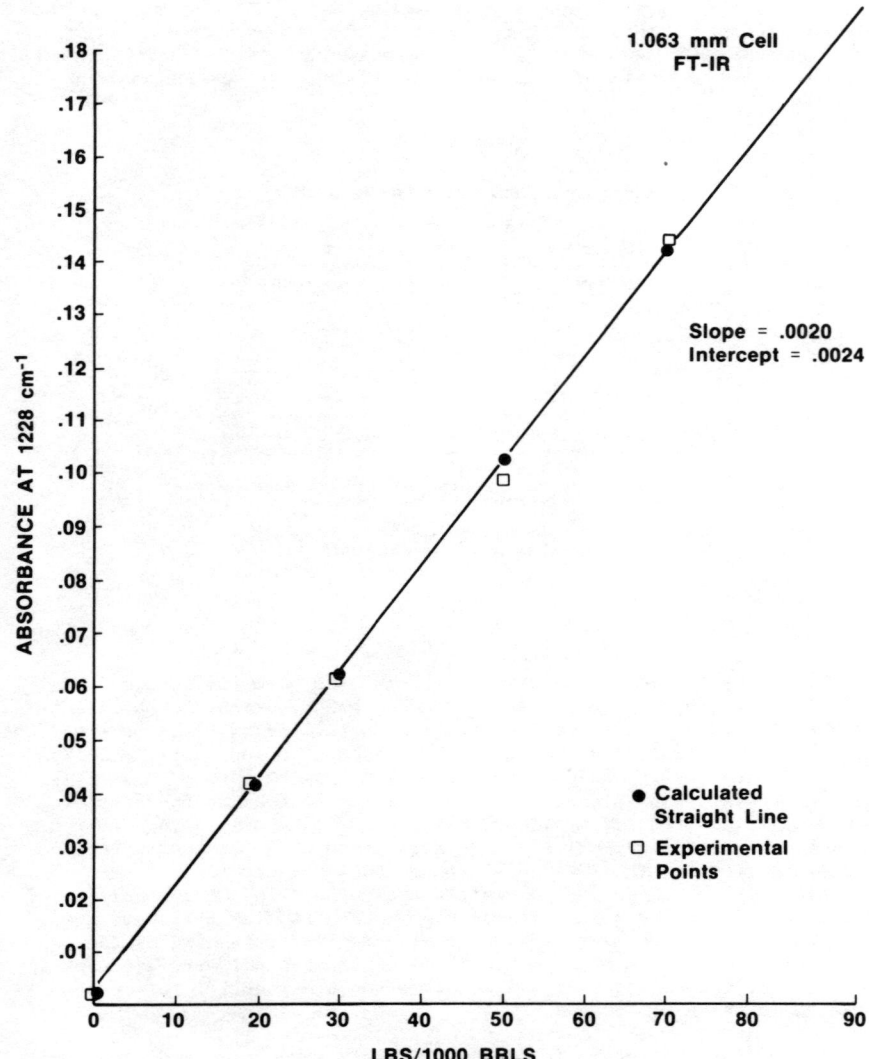

Figure 11   Calibration for additive analysis in gasoline

that has been made on quantitative applications of infrared through the introduction of computerized infrared instruments. The ease with which the spectroscopist can now program a quantitative analysis of complex multi-component determinations, or truly push the limits of detection in trace analysis is apparent in the flood of literature which has appeared in the last few years on such applications (G1-9). With the advent of low-cost and reliable FT-IR instruments, even quality control applications are proliferating. Figure 11 shows a typical calibration curve for the determination of gasoline additive in a fuel. The analysis is done at the terminal blending facility, and control limits can be established for the analysis in a feedback loop to the blending operation. An excellent example of trace anlysis using FT-IR was published by Mantz (G9), where less than 1 ppb of nickel carbonyl in the presence of CO was demonstrated. Haaland (G6) has used least squares methods to improve the sensitivity and precision of quantitative analysis of trace gases by FT-IR. An example is shown in Table IV.

TABLE IV

DETECTION LIMITS FOR TRACE GASES[a]

Gas	A Detection Limit[b] (ppm) (maximum absorption)	B Detection Limit[c] (ppm) (least squares method III)
CO	2.6	0.6
N_2O	1.5	0.2
CO_2	0.43	0.08

[a] Based on a 10-cm pathlength, total pressure of 640 Torr, and a 35 min measurement time for each sample and background spectra.

[b] Detection limit defined as the concentration of gas whose maximum absorption peak lies 3 σ above the baseline noise.

[c] Detection limits from column A changed by the improvement in detection factor calculated by least squares method III.

Conclusion

It is easy to see the maturity of FT-IR as a powerful tool in the industrial laboratory. At Battelle's Columbus Laboratories, Jakobsen and coworkers (I1,2) have been on the frontiers of utilizing FT-IR to study the kinetics of the adsorption of blood proteins onto various surfaces. These biomedical experiments are done in real-time and with a very rapid time scale. Extremely high signal-to-noise is required because the complex aqueous solutions require many spectral subtractions before the desired information is obtained. With a time resolution of 0.8 seconds, these workers are studying the small differences between successive layers as whole blood is adsorbed, and to profile the changes in the composition of the adsorbed layers as a function of time. Such studies will have great practical significance since they will allow an understanding of the coagulation or clotting mechanisms that can occur when polymers are used in heart valves, catheters, dialysis membranes or other artificial organs implanted in the human body.

Finally, another dramatic advance in FT-IR spectroscopy has come with the utilization of sophisticated mathematical techniques for the deconvolution of complex band shapes in the infrared spectrum to the resolution limit of the spectrometer. This work, originating in the laboratories of the National Research Council in Canada, is described in an elegant set of papers by Kauppinen et al. (B3,7-9) Figure 12 shows a comparison of the infrared spectrum of chlorobenzene as resolved through self-deconvolution and by first, second, third and fourth derivatives. The derivative spectra are somewhat more limited by signal-to-noise than the deconvoluted results. But it is certainly clear that with these powerful techniques of mathematical processing of our data, Fourier transform infrared spectroscopy is securely in place in the industrial laboratory (B4) for both qualitative, quantitative, sophisticated structural, and quality control work in the industrial laboratory.

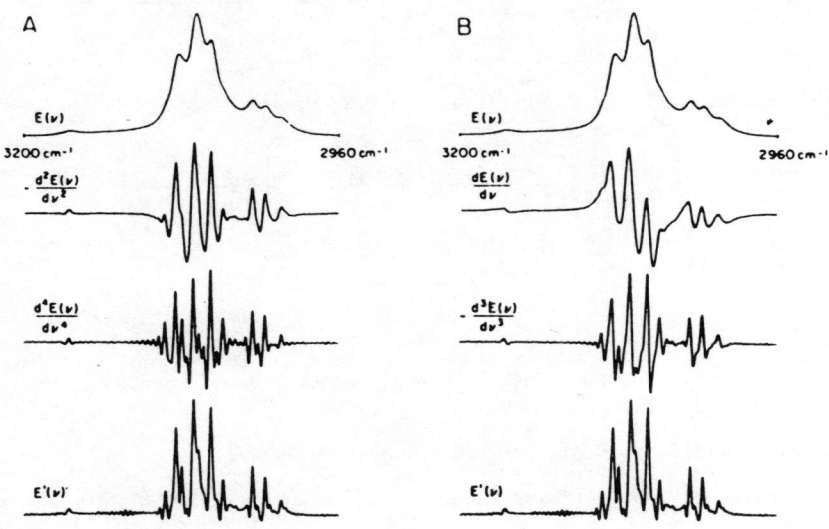

Figure 12   Comparison of the infrared spectrum $E(\nu)$, of chlorobenzene with the self-deconvoluted spectrum, $E'(\nu)$, and with the second and fourth (column A) and the first and third (column B) derivatives

Acknowledgements

I am totally indebted to the excellent Molecular Spectroscopy Group at Sohio, under the direction of Dr. J. Robert Mooney. In particular, I'd like to thank Dr. David Compton, Mr. Ron Kollar and Mrs. Patricia Wancheck for the specific examples I have utilized from our laboratories.

Bibliography

A. General

1. J.B. Bates, Science, 191 (4222), 31 (1976).

2. E.D. Becker and T.C. Farrar, Science, 178, 361 (1972).

3. J.A. deHaseth, "Fourier Transform Infrared Spectrometry", Ch. p. 387 in Fourier, Hadamard and Hilbert Transformations in Chemistry, ed. A.G. Marshall (Plenum Publ. Corp., 1982).

4. J.R. Ferraro and L.J. Basile, Eds., Fourier Transform Infrared Spectroscopy (Academic Press, Inc., New York, 1978).

5. J.G. Grasselli and L.E. Wolfram, Appl. Optics, 17, 1386 (1978).

6. J.G. Grasselli, P.R. Griffiths and R.W. Hannah, Appl. Spectrosc. 36:2 (1982).

7. P.R. Griffiths, Ed., Transform Techniques in Chemistry (Plenum Press, New York, 1975).

8. P.R. Griffiths, Chemical Infrared Fourier Transform Spectroscopy (Wiley, New York, 1978).

9. P.R. Griffiths, H.J. Sloane and R.W. Hannah, Appl. Spectrosc., 31, 485 (1977).

10. P.R. Griffiths, C.T. Foskett and R. Curbelo, Appl. Spectrosc. Rev., 6:1, 31 (1972).

11. P.R. Griffiths, Appl. Spectrosc., 31, 497 (1977).

12. R.O. Kagel and S.T. King, Ind. Res. (November, 1973).

13. J.L. Koenig, Am. Lab., 9 (September, 1974).

14. J.L. Koenig, Appl. Spectrosc., 29, 293 (1975).

15. J.L. Koenig and D.L. Tabb, Can. Res. and Devel. (September/October, 1974).

16. A.L. Smith, Applied Infrared Spectroscopy (Wiley, New York, 1979).

17. Spectral Collections

    a. Coblentz Society Publications, C.D. Craver, Ed.

    Desk Book of Infrared Spectra
    Gases & Vapors
    Halogenated Hydrocarbons
    Plasticizers & Other Additives
    Regulated Chemicals (in preparation)

    Available from Coblentz Society, Box 9952, Kirkwood, MO 63122

b. Vapors

   EPA Vapor Phase Library - Environmental Protection Agency
                            Washington, DC

   Sadtler Vapor Phase Library - Sadtler Reserch Laboratories
                            Philadelphia, PA

c. Liquids and Solids

   Aldrich/Nicolet Library of Infrared Spectra - Nicolet Analytical
                                                 Instruments
                                                 Madison, WI

   Sadtler Liquid and Solid FT-IR Library - Sadtler Research Labs
                                            Philadelphia, PA

d. Indices

   American Society for Testing and Materials
   distributed by Sadtler Labs., 3314 Spring Garden St.,
   Philadelphia, PA 19104 (1974).

   Comprehensive indices coded by Committee E-13.03 for the
   infrared spectra in all of the general collections above plus
   infrared spectra abstracted from technical journals through
   1972. These are available in two-volume sets for each type of
   index, with a total list of 145,000 spectra in each set.

   1. Molecular Formula List of Compounds, Names and References to
      Published Infrared Spectra
      AMD-31   92,000 compounds
      AMD-31-S15   53,000 compounds

   2. Serial Number List of Compounds, Names and References to
      Published Infrared Spectra
      AMD-32
      AMD-32-S15

   3. Alphabetical List of Compound Names, Formulae, and
      References to Published Infrared Spectra
      AMD-34
      AMD-34-S15

   Atlas of Spectral Data & Physical Constants for Organic
   Compounds
   2nd Edition, J.G. Grasselli and W.M. Ritchey, Editors, CRC
   Press, Inc., 18901 Cranwood Parkway, Cleveland, OH 44128
   (1975).

   Contains coded infrared spectra for 22,000 compounds. It lists
   strong bands in the infrared and includes Raman, UV, NMR and
   mass spectral data when available. Extensive cross-references
   and indices.

   Index of Vibrational Spectra of Inorganic and Organometallic
   Compounds, Vol. 1
   N.N. Greenwood, E.J.F. Foss and B.P. Straughan, CRC Press, Inc.,
   18901 Cranwood Parkway, Cleveland, OH 44128 (1972).

B. Data Processing

1. R.J. Anderson and P.R. Griffiths, Anal. Chem., 47, 2339 (1975).

2. R.J. Anderson and P.R. Griffiths, Anal. Chem., 50, 13, 1804 (1978).

3. D.G. Cameron, J.K. Kauppinen, D.J. Moffat and H.H. Mantsch, Appl. Spectrosc., 36, 245 (1982).

4. D.A.C. Compton and J.R. Mooney (The Standard Oil Company, Cleveland, OH), "Spectroscopic Uses for Fourier Self-deconvolution", Paper No. 661, Pittsburgh Conference, 1983.

5. R. Geick and Z. Fresenius, Anal. Chem., 88, 1 (1977).

6. G. Horlick, Anal. Chem., 43, 61A (1971).

7. J.K. Kauppinen, D.J. Moffatt, H.H. Mantsch and D.G. Cameron, Appl. Spectrosc., 35:3, 267 (1981).

8. J.K. Kauppinen, D.J. Moffatt, H.H. Mantsch and D.G. Cameron, Anal. Chem., 53, 1454 (1981).

9. J.K. Kauppinen, D.J. Moffatt, D.G. Cameron and H.H. Mantsch, Appl. Optics, 20, 1866 (1981), and Appl. Optics, 21, 1866 (1982).

C. Polymers

1. M.K. Antoon, K.M. Starkey and J.L. Koenig, "Applications of FT-IR Spectroscopy to Quality Control of the Epoxy Matrix", ASTM Spec. Tech. Publ. 1979, STP 674, pp. 541-52.

2. A. Chambles, "A Study of Polyurethane Cure by Infrared Spectrophotometry", Rep. 1977, Y/DK-183, Union Carbide, Oak Ridge, TN.

3. W.W. Hart, P.C. Painter, J.L. Koenig and M.M. Coleman, Appl. Spectrosc., 31, 220 (1977).

4. H. Ishida and J.L. Koenig, Am. Lab., 33 (March, 1978).

5. J.L. Koenig and M.K. Antoon, Appl. Optics, 17, 1374 (1978).

6. J.L. Koenig and D.L. Tabb, Canadian Res. and Dev., September/October, 1974.

7. P.C. Painter, M.M. Coleman and J.L. Koenig, The Theory of Vibrational Spectroscopy and its Application to Polymeric Materials (John Wiley & Sons, Inc., New York, 1982).

8. H.W. Siesler, Ch. "Characterization of Chemical and Physical Changes of Polymer Structure by Rapid-Scanning Fourier Transform IR (FT-IR)" (Practical Spectroscopy Series, ed. H. Siesler & K. Holland-Moritz, Vol. 4), Marcell Dekker, Inc., New York (1980).

9. J.F. Sprouse, B.M. Halpin, Jr., and R.E. Sacher, "Cure Analysis of Epoxy Composites Using Fourier Transform Infrared Spectroscopy", (AMMRC TR 78-45), Army Materials and Mechanics Research Center, Watertown, MA 02172 (DRXMR-RA).

D. Coal, Shale, Minerals and Heavy Hydrocarbons

1. W.W. Hart, P.C. Painter, J.L. Koenig and M.M. Coleman, Appl. Spectrosc., 31, 220 (1977).

2. P.C. Painter, S.M. Rimmer, R.W. Snyder and A. Davis, Appl. Spectrosc., 35, 102 (1981).

3. P.C. Painter, M.M. Coleman, R.W. Snyder, O. Majahan, M. Komatsu and P.L. Walker, Jr., Appl. Spectrosc., 35, 106 (1981).

4. P.C. Painter, R.W. Snyder, M. Starsinic, M.M. Coleman, D.W. Kuehn and A. Davis, Appl. Spectrosc., 35, 475 (1981).

5. P.C. Painter, R.W. Snyder, D.E. Pearson and J. Kwong, Fuel, 59, 282 (1980).

6. P.C. Painter, R.W. Snyder, J. Youtcheff, P.H. Given, H. Gong and N. Suhr, Fuel, 59, 364 (1980).

7. P.C. Painter, M.M. Coleman, R.G. Jenkins, P.W. Whang and P.L. Walker, Jr., Fuel, 57, 337 (1978).

8. P.C. Painter and M.M. Coleman, Fuel, 58, 301 (1979).

9. P.R. Solomon, Adv. Chem. Series, 192, Am. Chem. Soc., 95 (1981).

10. P.R. Solomon and F.P. Miknis, Fuel, 59, 893 (1980).

11. P.C. Uden, "Thermal Analysis and Pyrolysis Gas Chromatography of Oil Shale", Proc. 10th North American Thermal Analysis Conference, Boston, Mass., p. 233 (1980).

E. Combined Techniques

1. S. Bourne, G.T. Reedy and P.T. Cunningham, J. Chromatog. Science, 17, 460 (1979). (GC/Matrix Isolation/FT-IR)

2. D.A.C. Compton, J.G. Grasselli and M.L. Mittleman (The Standard Oil Company, Cleveland, OH), "Use of a Small FT-IR Spectrometer as a Mobile Detector for Fluid Streams", 1983 International Conference on Fourier Transform Infrared Spectroscopy, University of Durham, U.K.

3. M.D. Erickson, Appl. Spectrosc. Rev., 15, 261 (1979). (GC)

4. K.L. Gallaher and D.B. Lukco (The Standard Oil Company, Cleveland, OH), "GC/MS and GC/FT-IR; a Comparison", Am. Soc. Mass. Spectrosc. Mtg. (May, 1982).

5. S.E. Garlock, G.E. Adams and S.L. Smith, Am. Lab., 48 (December, 1982). (Capillary GC)

6. W. Heres, "GC-FTIR: Direct Elucidation of Substitution Patterns using Packed and Capillary Columns", FT-IR Application Note 15, Bruker Analytische Messtechnik GMBH, Rheinstetten, West Germany.

7. T. Hirschfeld, Anal. Chem., 52, 2997A (1980). (Hyphenated Techniques)

8. J. Chromatog. Science, Special Issue (August, 1979).

9. K. Krishnan, R.H. Brown, S.L. Hill, S.C. Simonoff, M.L. Olson and D. Kuehl, Am. Lab. (March, 1981). (Capillary GC)

10. D. Kuehl, G.J. Kemeny and P.R. Griffiths, Appl. Spectrosc., 34, 222 (1980). (Capillary GC)

11. D. Kuehl and P.R. Griffiths, Anal. Chem., 52, 1394 (1980). (LC)

12. J.O. Lephardt and R.A. Fenner, Appl. Spectrosc., 35, 95 (1981). (EGA)

13. J.O. Lephardt and R.A. Fenner, Appl. Spectrosc. 34:2, 174 (1980). (EGA)

14. J.O. Lephardt, Appl. Spectrosc. Rev., 18:2, 265 (1982-83). (EGA)

15. S.A. Liebman, D.H. Ahlstrom and P.R. Griffiths, Appl. Spectrosc., 30, 355 (1976). (EGA)

16. V. Rossiter, Am. Lab., 144 (February, 1982). (GC)

17. V. Rossiter, Am. Lab., 71 (June, 1982). (Capillary GC)

18. K.H. Shafer, S.V. Lucas and R.J. Jakobsen, J. Chromatog. Science, 17, 464 (1979). (Capillary GC)

19. K.H. Shafer, M. Cooke, F. DeRoos, R.J. Jakobsen, O. Rosario and J.D. Mulik, Appl. Spectrosc., 35, 469 (1981). (LC, GC, GC/MS)

20. D.W. Vidrine, J. Chromatog. Science, 17, 477 (1979). (LC)

F. Special Techniques (Reflectance, Photoacoustic, Matrix Isolation)

1. D.B. Chase, R.L. Amey and W.G. Holtje, Appl. Spectrosc., 36, 155 (1982). (DRIFT)

2. M.P. Fuller and P.R. Griffiths, Anal. Chem., 50, 1906 (1978). (DRIFT)

3. M.P. Fuller and P.R. Griffiths, Appl. Spectrosc., 34, 533 (1980). (DRIFT)

4. J.A. Gardella, Jr., E.M. Eyring, J.C. Klein, and M.B. Carvalho, Appl. Spectrosc., 36, 570 (1982). (PAS)

5. L.B. Lloyd, R.C. Yeates, and E.M. Eyring, Anal. Chem., 54, 549 (1982). (PAS)

6. J.B. Kinney and R.H. Staley, Anal. Chem., 55, 343 (1983). (PAS)

7. K. Krishnan, Appl. Spectrosc., 35, 549 (1981). (PAS)

8. D. Kuehl and P.R. Griffiths, Chromatog. Science, 17, 471 (1979). (DRIFT)

9. M.J.D. Low and G.A. Parodi, Appl. Spectrosc., 34, 76 (1980). (PAS)

10. S.R. Lowry, D.G. Mead and D.W. Vidrine, Anal. Chem., 54, 546 (1982). (PAS)

11. G. Mamantov, A.A. Garrison and E.L. Wehry, Appl. Spectrosc., 36:4, 339 (1982). (Matrix Isolation)

12. J.F. McClelland, Anal. Chem., 55, 89A (1983). (PAS)

13. M. Mehicic, R.G. Kollar and J.G. Grasselli, "Analytical Applications of Photoacoustic Spectroscopy using FT-IR", Proceedings of the International Conference on Fourier Transform Infrared Spectroscopy held at the University of South Carolina, 289, 99 (1981).

14. G.L. Powell, "Surface Analysis by Fourier Transform Infrared (FT-IR) Spectroscopy", Rep. 1981, Y/DU-192, Union Carbide Corporation, Oak Ridge, TN. (DRIFT)

15. S.M. Riseman, S.I. Yaniger, E.M. Eyring, D. Macinnes, A.J. Heeger and A.G. Macdiarmid, Appl. Spectrosc., 35, 557 (1981). (PAS)

16. M.G. Rockley, H.H. Richardson and D.M. Davis, J. Photoacoustics, 1:1, 145 (1982). (PAS)

17. A. Rosencwaig, Anal. Chem., 47, 592A (1975). (PAS)

18. D.W. Vidrine, Appl. Spectrosc., 34, 314 (1980). (PAS)

19. E.L. Wehry and G. Mamantov, Anal. Chem., 51, 643A (1979). (Matrix Isolation)

G. Quantitative and Trace Analysis

1. M.K. Antoon, J.H. Koenig, and J.L. Koenig, Appl. Spectrosc., 31, 518 (1977).

2. C.W. Brown, P.F. Lynch and R.J. Obremski, Appl. Spectrosc., 36, 539 (1982).

3. R. Cournoyer, J.C. Shearer and D.H. Anderson, Anal. Chem., 49, 2275 (1977).

4. H.E. Diem and S. Krimm, Appl. Spectrosc., 35, 421 (1981).

5. P.C. Gillette and J.L. Koenig, Appl. Spectrosc., 36, 535 (1982).

6. D.H. Haaland and R.G. Easterling, Appl. Spectrosc., 34, 539 (1980).

7. T. Hirschfeld, Anal. Chem., 48, 721 (1976).

8. J.L. Koenig and D. Kormos, Appl. Spectrosc., 33, 349 (1979).

9. A.W. Mantz, Appl. Spectrosc., 30, 539 (1976).

H. Inorganic Analysis

1. R.M. Gendreau, R.J. Jakobsen, W.M. Henry and K.T. Knapp, Journal of E.S.&T., 14, 990 (August, 1980).

I. Biomedical and Pharmaceutical

1. R.M. Gendreau, S. Winters, R.I. Leininger, D. Fink, C.R. Hassler and R.J. Jakobsen, Appl. Spectrosc., 35:4, 353 (1981).

2. R.M. Gendreau, Appl. Spectrosc., 36:1, 47 (1982).

3. J.A. Ryan, Am. Lab (Nov. 1981).

4. P.H.G. Van Kasteren, Ch. Analytical Applications of FT-IR to Molecular and Biological Systems, ed., J.R. Durig (Dordrecht, 1980).

J. **Interesting Novel Applications**

1. D.G. Cameron, J. Umemura, P.T. Wong and H.H. Mantsch, Colloids & Surfaces, 4, 131 (1982). (Micelles)

2. R.J. Jakobsen, C.J. Riggle and E.J. Drauglis, Appl. Spectrosc., 36, 570 (1982). (Solid/Liquid Interfaces)

3. G.J. Kemeny and P.R. Griffiths, "Feasibility of Using Dual-beam FT-IR Spectrometry to Study Adhesives or Metal Surfaces", Private Communications.

4. J. Umemura, H.H. Mantsch and D.G. Cameron, J. Colloid and Interface Science, 83, 558 (1981). (Micelles)

5. J. Umemura, D.G. Cameron and H.H. Mantsch, J. Phys. Chem., 84:18, 2272 (1980). (Micelles)

FT-IR SPECTRA OF COORDINATION COMPOUNDS

Ian S. BUTLER, Jacqueline SEDMAN, and Ashraf A. ISMAIL

Department of Chemistry, McGill University,
801 Sherbrooke St. West, Montreal,
Quebec, Canada, H3A 2K6.

As in almost every other area of application, FT-IR spectroscopy has proven to be a benefit to inorganic chemists as well. The importance of high resolution, rapid spectral scans, and the ability to work with small quantities of material by co-adding spectra have all been exploited. In addition, the computer subtraction facet of FT-IR has greatly increased the potential of this technique in inorganic chemistry. In this presentation, some typical examples of the various applications in coordination chemistry will be demonstrated. The reader is referred to the literature for a recent review on the application of FT-IR in matrix isolation studies, e.g., uranium oxides (UO, UO_2, and UO_3), xenon fluorides (XeF, XeF_2), and germanium halides [1]. Although not as yet examined, the potential of FT-IR spectroscopy in identifying metal-ligand vibrations in the far-infrared spectra of coordination complexes by studying the metal isotope shifts is clearly an important future development. Again, the reader is referred to the excellent review by Nakamoto [2] on the recent progress in inorganic vibrational spectroscopy for a summary of the metal isotope effect.

One area of interest concerns the assignment of the metal ligand vibrations in simple complexes through a comparison of the spectra of similar compounds. Often the observed data can be used to predict the geometrical structures of isomers although in the absence of supporting Raman data, the structures must be considered tentative. As an example of the FT-IR spectra of coordination compounds, we will consider the 1:1 addition compounds formed by triphenyltin(IV) bromide and iodide (Ph_3SnX, X = Br, I) with dimethyl-propylene-urea (DMPU) [3]. The bromo complex exhibits three characteristic

$$Ph_3SnX + DMPU \longrightarrow Ph_3SnX \cdot DMPU$$

bands in its far-infrared spectrum in Nujol mull (Fig. 1) at 278, 235 and 172 cm^{-1} which have been attributed to $\nu(Sn-Ph)$, $\nu(Sn-O)$, and $\nu(Sn-Br)$, respectively. Similarly, the bands at 276, 230, and 131 cm^{-1} have been assigned as $\nu(Sn-Ph)$, $\nu(Sn-O)$, and $\nu(Sn-I)$ in Fig. 2. The actual disposition of the five ligands in the presumably trigonal bipyramidal complexes cannot be elucidated on the basis of infrared data alone - Raman and nuclear magnetic resonance (^{13}C) will be needed to resolve the problem fully. However, these far-infrared data are typical of those obtainable fairly easily on an FT-IR spectrometer; all that is really needed is change the beamsplitter to a 12 micron Mylar film and then to run about 400 spectra, a matter of perhaps 15 minutes.

On a totally different front, we have recently completed some work on the application of FT-IR spectroscopy in the assignment of the $\nu(CO)$ modes in metal carbonyls [e.g., $M_2(CO)_{10}$, M = Mn, Re] through the use of nematic liquid crystals and polarized infrared radiation. Such $\nu(CO)$ modes occur in the neighbourhood of 2000 cm^{-1}, in a region relatively free from the absorptions of the liquid crystal solvents. Fig. 3 depicts the FT-IR (polarized parallel and perpendicular) of $Mn_2(CO)_{10}$ dissolved in a room temperature liquid crystal

(Merck 1132) [4]. For details on the preparation of oriented liquid crystal solutions, see ref. 5. The spectra in Fig. 3 clearly show the effect of changing the polarization plane of the incident infrared radiation and since $Mn_2(CO)_{10}$ possesses D_{4d} molecular symmetry, the B_2 $\nu(CO)$ modes are expected to be polarized parallel, while the E_1 mode should be polarized perpendicular. This situation is exactly what is observed in the subtracted spectrum shown in Fig. 4. The same results have been reported by Gray et al. [6] using a different liquid crystal solvent and a dispersive infrared instrument. However, the facility of spectral subtraction with the FT-IR instrument renders such polarization measurements extremely easy.

We will now turn our attention to another example of the application of FT-IR, this time in the area of hormone-receptor site detection through the detection of organometallic labels [7]. When an estrogen molecule enters a cell from the blood stream it searches for a specific receptor protein and the resultant estrogen-receptor complex ultimately affects new protein synthesis in the cell nucleus.

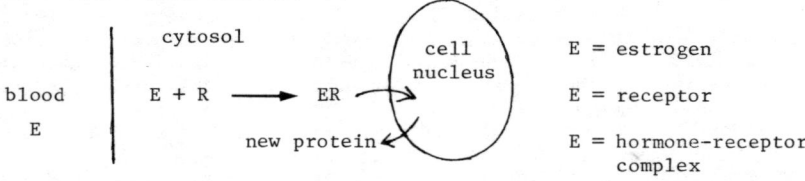

E = estrogen
E = receptor
E = hormone-receptor complex

The interaction of estrogens with their specific hormone receptors has been the focus of much research [8] since it is becoming clear that the process is intimately involved in certain hormone-dependent cancers (e.g., breast cancer). There appears to be a relationship between the hormone receptor site concentration and the onset of cancer - prior to breast cancer for instance there is apparently an increase in the hormone-receptor site concentration. These concentrations are monitored already in several ways, especially radiochemical techniques, e.g., tritium-labelled estradiol. Our approach has been radically different in that we have initially formed organochromium-labelled species, such as those shown below, by well-established synthetic routes [9]. Then, these complexes have been incubated

See reaction schemes 1 and 2.

with the protein extracted from lamb uterus prior to protein precipitation. By comparing the FT-IR spectra of the "free" protein and the labelled species it is relatively easy to detect the $Cr(CO)_3$ and $Cr(CO)_2(CS)$ labels by the appearance of the $\nu(CO)$ modes at around 2000 cm^{-1}. Again, notice that the labels were deliberately chosen so as to exhibit strong absorptions in the "window" of the protein spectra (Figs. 5-10).

The next step in this research will be to attempt to quantify the technique in order to compare the results with those from the radiochemical measurements. However, we have already been able to obtain an excellent correlation between the weight of the sample and the area under the higher energy peak in the $\nu(CO)$ region (Fig. 11).

References

1. D.W. Green and G.T. Reedy in "Fourier Transform Infrared Spectroscopy - Applications to Chemical Systems", Vol. I, eds. J.R. Ferraro and L.J. Basile, Academic Press Inc., New York, 1978, pp. 1-59.

2. K. Nakamoto, J. Spectrosc. Soc. Japan, 30, 437 (1981)
3. C. Aitken and M. Onysczhuk, unpublished results.
4. J. Sedman and I.S. Butler, unpublished results.
5. B.J. Bulkin and K. Brezinsky, J. Chem. Phys., 15, 69 (1978).
6. R.A. Levenson, H.B. Gray and G.P. Caesar, J. Amer. Cham. Soc., 92, 3653 (1970).
7. A.A. Ismail, S. Top, A. Vessières, G. Jaouen and I.S. Butler, unpublished results.
8. E.V. Jensen and E.R. DeSombre, Science, 182, 126 (1973).
9. (a) G. Jaouen and G. Simonneaux, Inorg. Synth., 19, 197 (1979); (b) I.S. Butler, Acc. Chem. Res., 10, 359 (1977); (c) G. Pouskouleli, I.S. Butler and J.P. Hickey, J. Inorg. Nucl. Chem., 42, 1659 (1980).

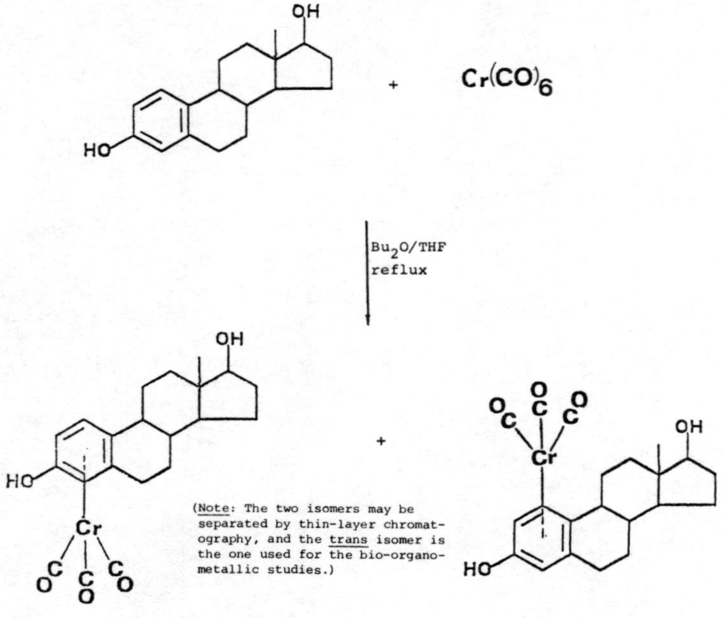

Scheme 1

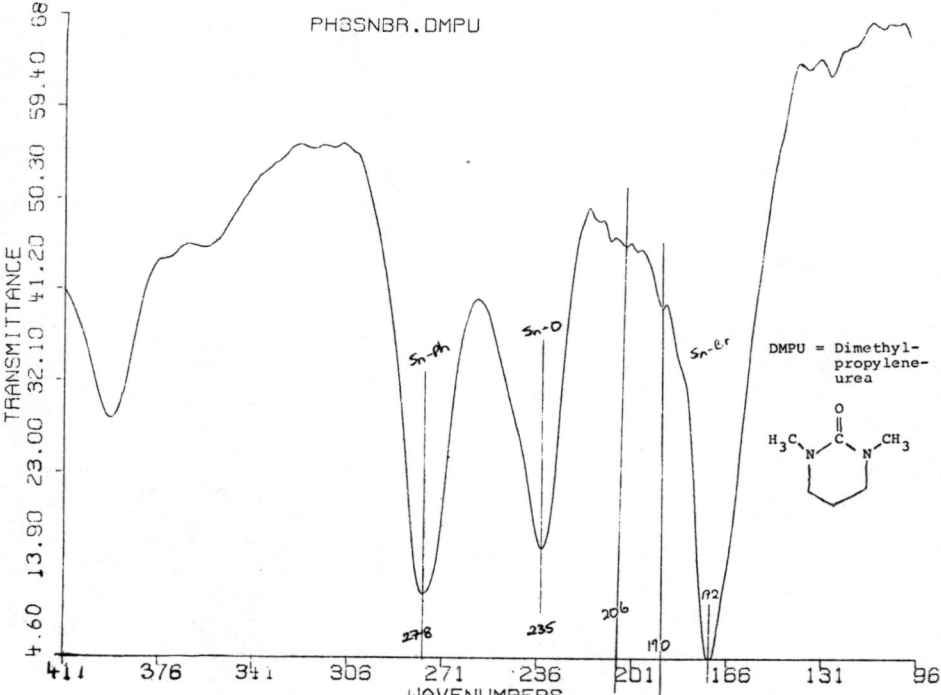

Fig. 1. Far-IR spectrum of Ph$_3$SnBr·DMPU (Nujol mull); 400 scans, 4 cm^{-1} resolution, ratioed against a Nujol background. (<u>Note</u>: All FT-IR spectra were recorded on a Nicolet model 6000 spectrometer equipped with a mercury-cadmium-telluride (MCT), liquid-nitrogen cooled detector).

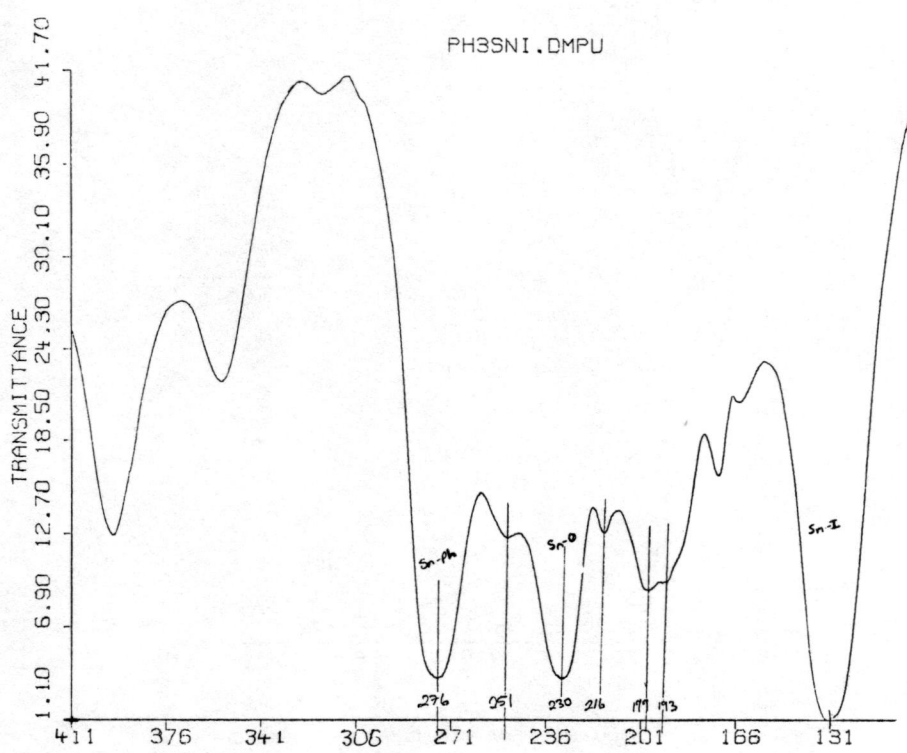

Fig. 2. Far-IR spectrum of Ph₃SnI·DMPU (same conditions as Fig. 1).

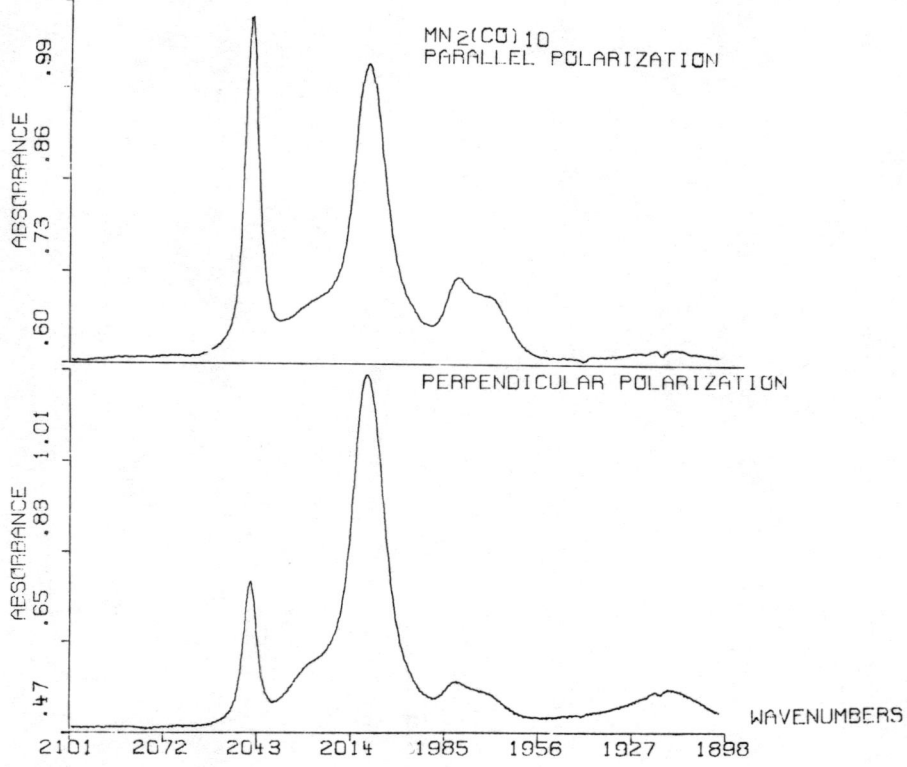

Fig. 3. FT-IR in the $\nu(CO)$ region of $Mn_2(CO)_{10}$ dissolved in Merck 1132 liquid nematic liquid crystal using both parallel and perpendicularly polarized IR radiation; 200 scans, 4 cm^{-1} resolution, ratioed against a dry, CO_2-free, air background.

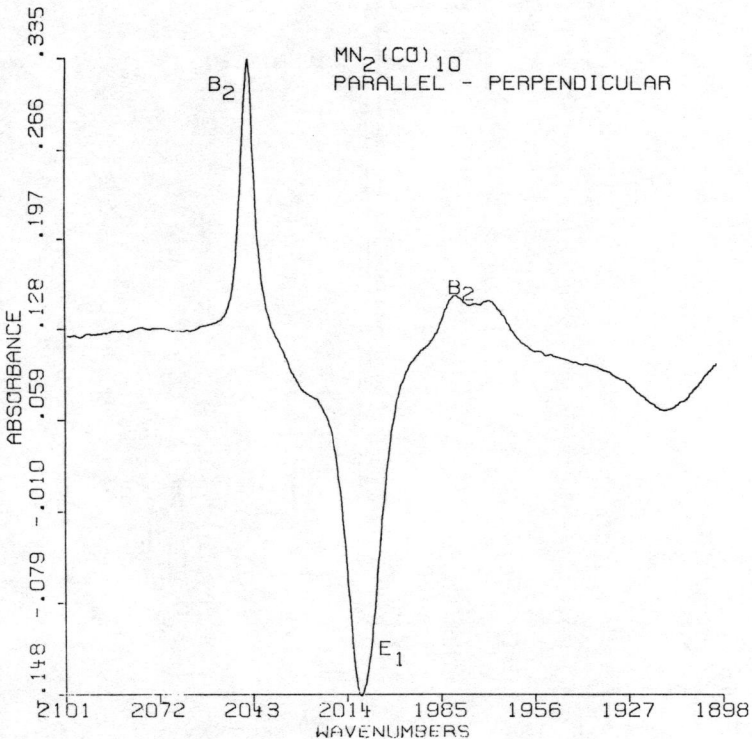

Fig. 4. Result of subtracting the perpendicular spectrum from the parallel one in Fig. 3.

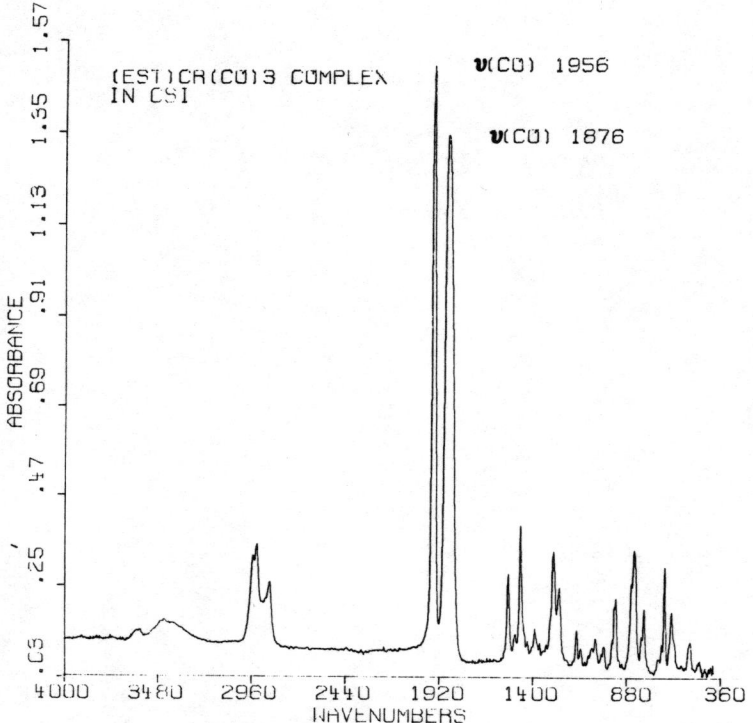

Fig. 5. FT-IR spectrum (4000-400 cm^{-1}) of (η^6-estradiol)Cr(CO)$_3$ (CsI disk); 32 scans, 4 cm^{-1} resolution, ratioed against a dry, CO$_2$-free, air background.

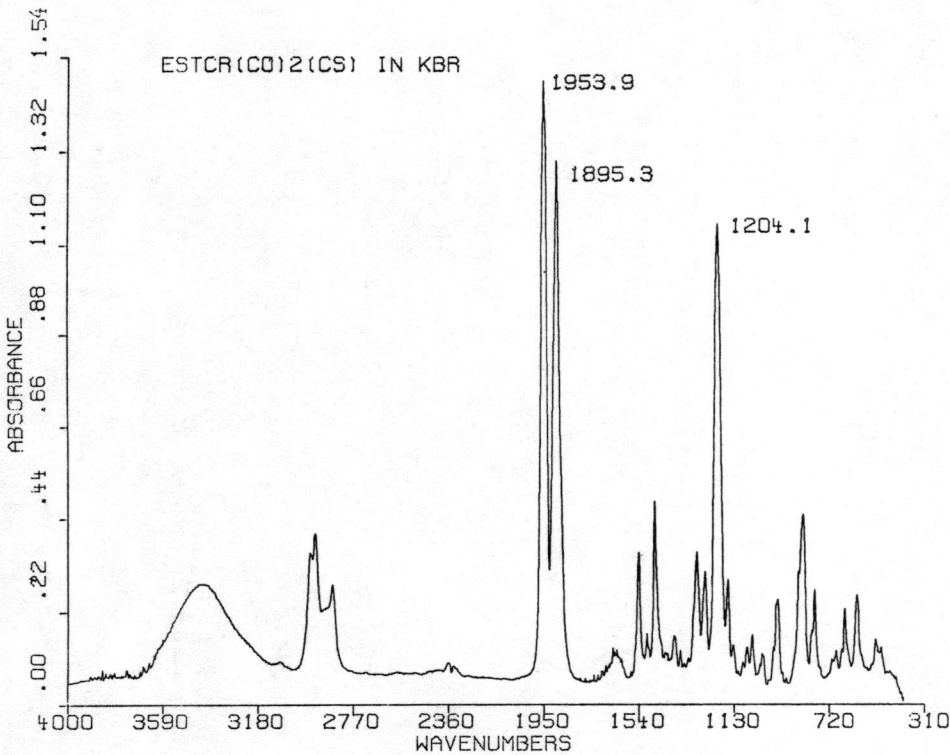

Fig. 6. FT-IR of (η^6-estradiol)Cr(CO)$_2$(CS) (same conditions as Fig. 5).

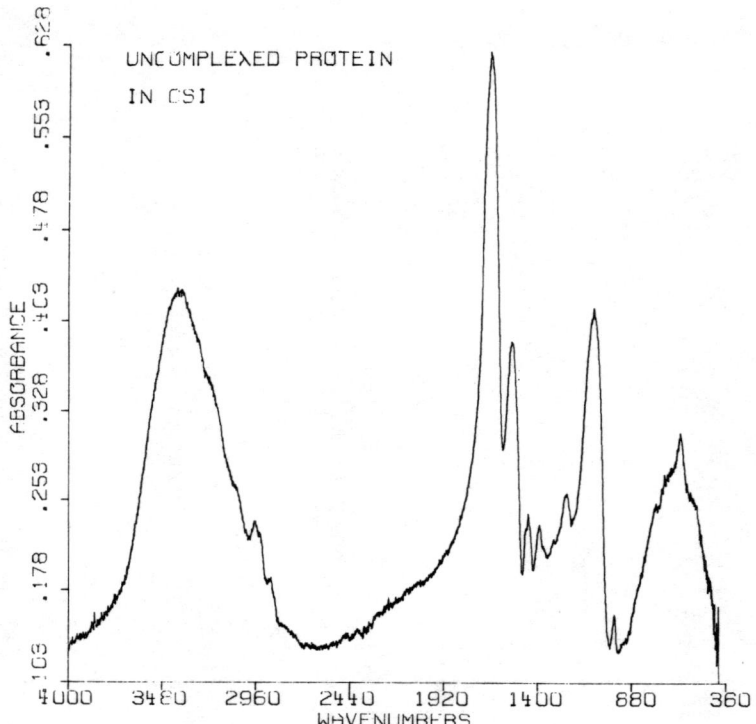

Fig. 7. FT-IR spectrum (4000-400 cm^{-1}) of the protein extracted from lamb uterus (CsI disk); 32 scans, 4 cm^{-1} resolution, dry, CO_2-free, air background.

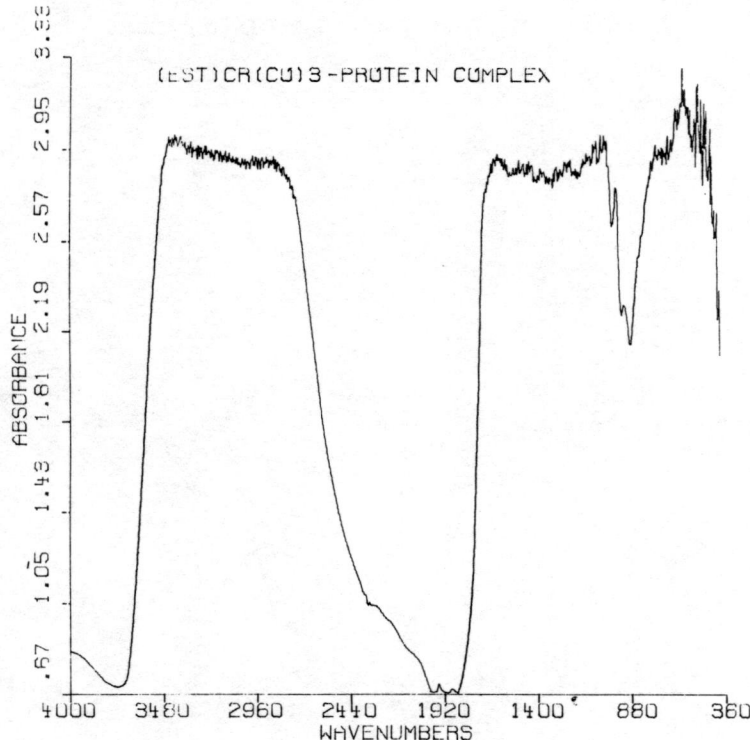

Fig. 8. Difference spectrum of (protein extract + metal carbonyl label) minus spectrum of pure protein; about 1.5 mg of sample pressed into a 3-mm mini-pellet, 20,000 scans, 4 cm^{-1} resolution, dry (CO_2-free) air background.

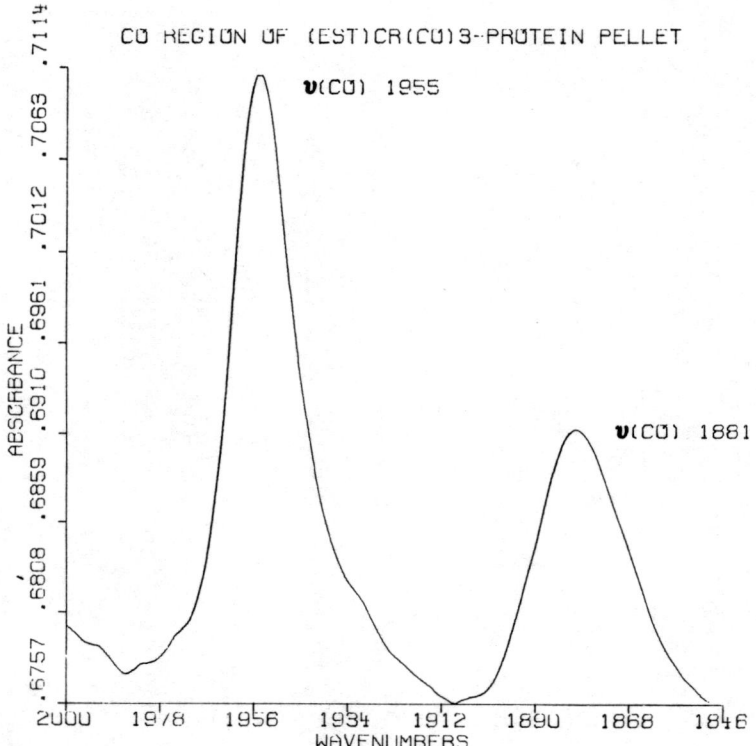

Fig. 9. ν(CO) region (expansion) from Fig. 8.

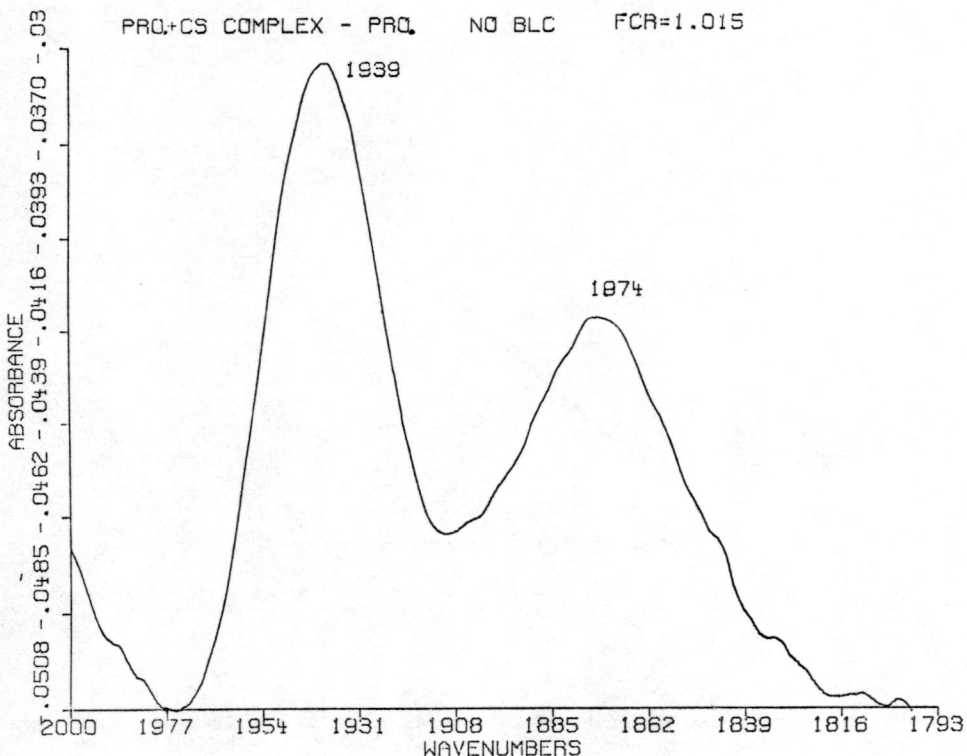

Fig. 10. ν(CO) bands detected for the same experiment but using (η^6-estradiol)$Cr(CO)_2(CS)$ as the labelling reagent for the protein extract.

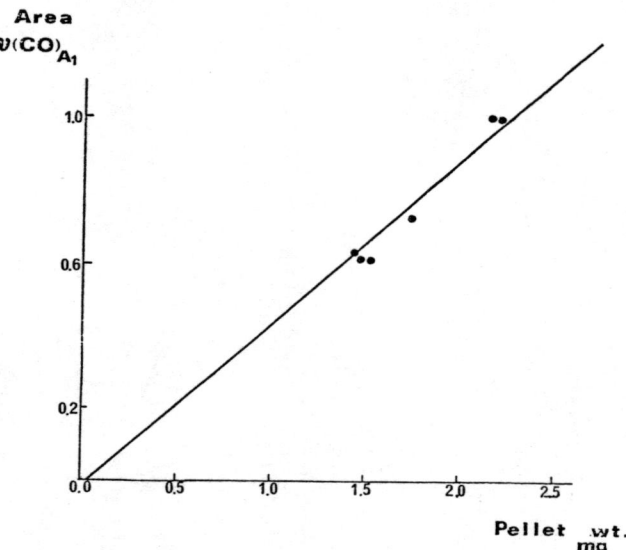

Fig. 11. Correlation of weight of the protein-labelled pellet versus the area of the higher-energy $\nu(CO)$ band.

METAL CLUSTERS AS MODELS OF METAL SURFACES - SOME VIBRATIONAL ASPECTS

S. F. A. KETTLE
School of Chemical Sciences, University of East Anglia,
Norwich, NR4 7TJ, U.K.

In recent years there has been an explosive growth in the study of transition metal cluster compounds. Many factors have contributed to this growth - the development of preparative and separative procedures, the expectation that cluster compounds mimic catalytic metal surfaces and so may provide insights into catalytic mechanisms (as well as, potentially, being catalysts themselves), the development of new techniques by which metal surfaces - and species absorbed onto them - may be studied, the recognition that novel bonding patterns may exist within them, and their involvement in bioinorganic chemistry, particularly as electron sinks and sources.

The extent to which the aggregate of metal atoms in a cluster mimic a metal surface is an issue far from settled. Frequently, clusters contain a metal cage which is, essentially, a fragment of the metal. Equally, however, some cages of metal atoms seem only to exist because they are 'sewn together' by bridgings ligands, with little or no direct metal-metal bonding. A continuum almost seems to exist between the two. From the other end, there have been many studies of the relationships between the properties of aggregates of bare metal atoms with the number of, and arrangement of, atoms within the aggregate.[1-12]

At U.E.A. there is a group of chemists who, for several years, have made a study of metal and other surfaces and species absorbed on them (Norman Sheppard and Mike Chesters) - using F.T.I.R., Raman, LEED and EELS methods. Another group (including Don Powell and Ian Oxton) has been more concerned with inorganic clusters using IR and Raman methods but the overlap between these two groups has been so great that it is often meaningless to distinguish between them. Our work has been carried out in close collaboration with the Lewis/Johnson

group at Cambridge and, in some parts the Stanghellini group
in Turin. In this paper I summarise some of our findings.

Before turning to the problem of ligands, I consider the
metal atoms themselves: how may we describe their vibrations?
If we can describe them effectively we shall, hopefully,
become aware of occasions when the metal atoms are bonded
unusually, or coupled unusually, because our 'rules' would
then not apply.

The most useful approach to metal atom vibrations seems
to be the 'plastic cluster' model introduced by Spiro.[13] In
this, one assumes that the stretching vibrations of one
metal-metal bond has no electronic awareness of whether
adjacent metal-metal bonds are stretching, contracting or are
static. The sole coupling between the motion of adjacent bonds
is mechanical in origin. That is, the coupling appears in the
G matrix, not the F. In the model, the vibrations of any
ligands are ignored. It is difficult to totally justify this
neglect - the fact that 'metal-metal' frequencies are rather
lower than most ligand motions is the best reason. However,
for whatever reason the model gives remarkably good agreement
with experimental data for new compounds[14] and, since it
contains few parameters, is capable of giving surprisingly good
bond length and bond angle predictions.[15] If interaction
constants are to be added to the model, it seems that they may
be uniformally set as -8% of the principal force constants for
transition metal clusters.[16]

The next step is to introduce ligands into a cluster, and,
correspondingly, adatoms or ligands on a surface. It seems
that vibrational spectroscopy may enable the chemical nature
of surface species (often in doubt) to be determined.[17]

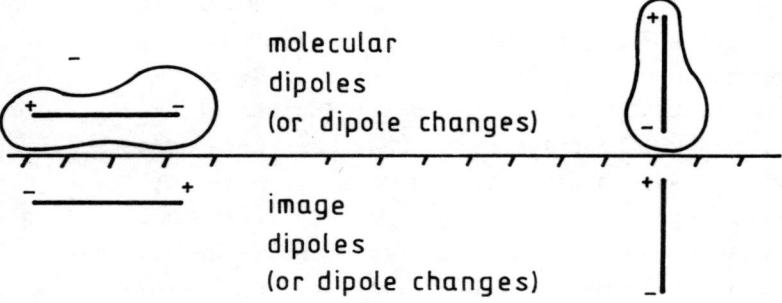

Two questions at once arise: which vibrations of ligands on metal surfaces will be seen in I.R.? To what extent are the vibrations of the ligands independent of one another? The answer to the first question is easy: there is a so-called 'surface selection rule' such that only vibrations with a dipole perpendicular to the metal surface are infrared active.[18] This is because a dipole moment change in an absorbed molecule induces an opposed dipole in the metal. This opposition leads to a zero nett dipole for dipoles parallel to the metal surface but an enhanced dipole perpendicular. (See the diagram). This selection rule is also applicable to specular EELS (electron energy loss spectroscopy) measurements but is relaxed in off-specular spectra (specular: the angle of study of the reflected electron beam equals the angle of the incident electron beam onto the surface).[19]

There has been much discussion of the vibrational interaction between surface species. It seems that such interactions often occur and manifest themselves as shifts in spectral bands.[20-22] Multiplicity of spectral bands does not seem to result. Rather, this phenomenon is to be associated with the occupancy of several different surface sites.[17] Yet life may be even more difficult than this.

The problem is well illistrated by clusters alone. There has recently been interest in the vibrational characteristics of the carbon atom in carbido-clusters. It seems that the vibration of this atom is almost entirely uncoupled from those of the metal atoms which surround it for the associated bands display the expected $\sqrt{\frac{12}{13}}$ frequency shift on ^{13}C substitution. Data are collected in the Table. Ignoring the last three species given, the vibrational patterns clearly reflect the symmetry of the local environment. To some extent the centre-of-gravity of the $\nu(M-C)$ frequencies reflects the overall charge on the species, but the pattern is not complete. The size of the 'hole' in the cluster occupied by the carbon atom and the local charge distribution around it may well be additional factors.

The last three examples given in the Table are of molecules containing a carbon atom on a surface (it is slightly above, and at the centre of, a square array of metal atoms). The motions of these carbon atoms is much more sensitive to the crystal environment than are those of the encapsulated carbon atoms but, from a band count and their frequencies, what is there to distinguish them from the species given above, particularly when it is recognised that low temperature spectra of the other species often show a splitting of degenerate modes? For monatomic ligands, at least, it may not always be easy to decide solely from vibrational data whether an atom is an adatom on a surface or whether it is incorporated into the lattice, close to the surface.

Now for some good news. It has been recognised[23] that for a

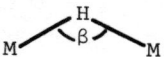

bridge the ratio of symmetric and antisymmetric $\nu(M-H)$ frequencies is given to a first approximation by

$$2^{\frac{1}{2}} \nu(\text{asym}) = \tan\beta \, \nu(\text{sym}).$$

For a hydrogen centring the face subtended by an equilateral triangle of metal atoms (M) one obtains[24] the relationship

$$2^{\frac{1}{2}}(1 + 3m_M \cos^2\beta) \, \nu(\text{asym}) = (1+3m_M \sin^2\beta) \, \nu(\text{sym}).$$

Such simple relationships have already enabled a rationalization of vibrational data for hydrogen absorbed on metal surfaces.[24,25]

Similarly, a detailed vibrational study of clusters containing a $C-CH_3$ (or $C-CD_3$) group bonded to an equilateral triangle of cobalt atoms in $Co_3(CO)_9C-X$ species and comparison with the EELS of platinum surfaces onto which either C_2H_2 or C_2H_4 has been absorbed, has shown that the latter both give rise to $C.CH_3$ species on the surface.[26]

Very similar has been the study of $Co_3(CO)_9CH$ and $Co_3(CO)_9CD$ which has shown that the adsorption of C_2H_2 on Ni(III) planes gives rise to CH groups on the surface.[27]

Another example is provided by clusters containing a μ_2 CH_2 group which have enabled an assessment of data reported for alleged μ_2 CH_2 groups on metal surfaces.[28]

In all of the cases I have given above it seems that the vibrational units are uncoupled one from another. As I have

TABLE

Species	C atom Site symmetry	Frequencies cm^{-1}	Ref
$Ru_6C(CO)_{17}$	ca O_h	838 (T_{1u})	a
$[Ru_6C(CO)_{16}]^{2-}$	distorted O_h	719 (E), 690 (A)	a
$[Co_6C(CO)_{15}]^{2-}$	D_{3d}	772 (A_2''), 719 (E')	b
$[Rh_6C(CO)_{15}]^{2-}$	D_{3d}	689 (A_2''), 653 (E')	b
$Co_6C(CO)_{12}S_2$	D_{3d}	819 (A_2''), 548 (E')	c
$[Os_{10}C(CO)_{24}]^{2-}$	O_h	753 (T_{1u})	d
$H_2Os_{10}C(CO)_{24}$	low?	773, 760, 735	d
$[Fe_6C(CO)_{16}]^{2-}$	C_s	818, 790, 776	a
$Fe_5C(CO)_{15}$	C_{4v}	805 (a_1), 775, 776 (e)	e
$Ru_5C(CO)_{15}$*	C_{4v}	757, 738, 730	e
$Os_5C(CO)_{15}$*	C_{4v}	793, 769, 757	e

* there are two crystallographically distinct molecules in the crystal structure.

a) P.L. Stanghellini, L. Cognolato, G. Bor and S.F.A. Kettle, Submitted for publication.
b) J.A. Creighton, R.D. Perfola, B.T. Heaton, S. Martinengo, L. Strona and D.A. Willis, J.Chem.Soc.Chem.Comm (1982) 864.
c) G. Bor and P.L. Stanghellini, J.Chem.Soc.Chem.Comm. (1979) 886.
d) I.A. Oxton, S.F.A. Kettle, P.F. Jackson, B.F.G. Johnson and J. Lewis. J.Mol.Struct., 71 (1981) 117.
e) I.A. Oxton, D.B. Powell, R.J. Goudsmit, B.F.G. Johnson, J. Lewis, W.J.H. Nelson, J.N. Nicholls, M.J. Rosales, M.D. Vargas and K.H. Whitmire, Inorg.Chim.Acta Let., 64 (1982) L259.

mentioned, this is not always the case. Carbon monoxide is a case in point, but, encouragingly, some respectable normal coordinate analyses are appearing for high-symmetry species,[29-31] although large carbonyl clusters remain a closed book and the extension of the data to metal surfaces an unexplored area. In the clusters, terminal CO groups on different metal atoms are coupled together, although, to a first approximation, the low frequency vibrations of CO groups are not.[32,33] It is difficult to escape the conclusion that the coupling mechanism is largely through-space and it is this recognition, rather than detailed aspects of the analysis of $\nu(CO)$ vibrations of metal clusters, which is important for surface chemistry.[22]

REFERENCES

1. H. Basch, J.Amer.Chem.Soc., 103 (1981) 4157.

2. J.J. Burton, J.Chem.Phys., 56 (1972) 3133.

3. J. Demuynck, M-M. Rohmer, A. Strick and A. Veillard, J.Chem.Phys., 75 (1981) 3443.

4. R.G. Wooley, Mol.Phys., 40 (1980) 381.

5. M. Gillet, Surface Science 67 (1977) 139.

6. J.F. Hamilton and R.C. Baetzold, Science, 205 (1979) 1213

7. H. Basch, M.D. Newton and J.W. Muskowitz, J.Chem.Phys., 73 (1980) 4492.

8. R.C. Baetzold, Inorg.Chem., 20 (1981) 118.

9. A.C. Farragher, Adv.Coll.' Interface Sci., 11 (1979) 3.

10. O. Sinanoğlu, Chem.Phys.Lett., 81 (1981) 188.

11. A.B. Anderson, J.Chem.Phys., 64 (1976) 4046.

12. M.R. Hoare and P. Pal, Adv.Phys., 20 (1971) 161.

13. T.G. Spiro, Prog.Inorg.Chem., 11 (1970) 1.

14. S.F.A. Kettle and P.L. Stanghellini, Inorg.Chem., 18 (1979) 2749.

15. S.F.A. Kettle and P.L. Stanghellini, Inorg.Chem., 21 (1982) 1447.

16. I.A. Oxton, Inorg.Chem., 19 (1980) 2825.

17. M.A. Chesters and N. Sheppard, Chem.in Brit., 17 (1981) 521.
18. H.A. Oearce and N. Sheppard, Surface Science 59 (1976) 205.
19. R.F. Willis "Vibrational Spectroscopy of Absorbates" Springer-Verlag, Berlin, 1980.
20. E.E. Mola, Surface Science 91 (1980) L45.
21 H. Metui and W.E. Palke, J.Chem.Phys., 69 (1978) 2574.
22. P. Hollins and J. Pritchard, Chem.Phys.Lett., 75 (1980) 378.
23. M.W. Howard, U.A. Jayasooriya, S.F.A. Kettle, D.B. Powell and N. Sheppard, J.Chem.Soc.Chem.Comm., (1979) 18. See also V. Baran, Inorg.Nucl.Chem.Lett., 6 (1970) 375.
24. J.A. Andrews, U.A. Jayasooriya, I.A. Oxton, D.B. Powell, N. Sheppard, P.F. Jackson, B.F.G. Johnson and J. Lewis, Inorg.Chem., 19 (1980) 3033.
25. U.A. Jayasooriya, M.A. Chesters, M.W. Howard, S.F.A. Kettle, D.B. Powell and N. Sheppard, Surface Science 93 (1980) 526.
26. P. Skinner, M.W. Howard, I.A. Oxton, S.F.A. Kettle, D.B. Powell and N. Sheppard, J.Chem.Soc. Faraday 2 77 (1981) 1203.
27. M.W. Howard, S.F.A. Kettle, I.A. Oxton, D.B. Powell, N. Sheppard and P. Skinner, J.Chem.Soc. Faraday 2 77 (1981) 397.
28. I.A. Oxton, D.B. Powell, N. Sheppard, K. Burgess, B.F.G. Johnson and J. Lewis, J.Chem.Soc.Chem.Comm., (1982) 719.
29. G.A. Battiston, G. Bor, U.K. Dietler, S.F.A. Kettle, R. Rossetti, G. Sbrignadello and P.L. Stanghellini, Inorg.Chem., 19 (1980) 1961.
30. G. Bor, G. Sbrignadello and F. Marcati, J.Organomet. Chem., 46 (1972) 357.

31. G. Bor, G. Sbrignadello and K. Noack, Helv.Chim.Acta, $\underline{58}$ (1975) 815.

32. S.F.A. Kettle and P.L. Stanghellini, J.Chem.Soc. Dalton (1982) 1175.

33. S.F.A. Kettle and P.L. Stanghellini. Unpublished work.

FT-IR SPECTRA OF NUCLEIC ACIDS AND
THE EFFECT OF METAL IONS

Theophile THEOPHANIDES

Université de Montréal, Département de Chimie
C.P. 6210, Succ. A, Montréal, Québec, H3C 3V1

CONTENTS

I. Introduction

II. Nomenclature

III. Structural Features of Nucleic Acid Constituents

IV. Metal Ions

V. Sample Handling

VI. Frequencies
- CH, CH_2, NH and OH Stretching Frequencies
- CO, C=N, C=C Stretching and NH_2 bending Frequencies
- Phosphate Frequencies
- Sugar Frequencies

VII. Effect of Metal Ions on Frequencies and Intensities

VIII. Phosphate Frequencies

I. Introduction

In this lecture we will be considering the interactions of metal ions with nucleic acids in view of the dramatic advances currently made towards the elucidation of the structure of nucleic acids (primary, secondary, tertiary) and the important role played by metal ions. Powerful new spectroscopic physicochemical techniques and X-ray crystallography are revealing new routes of thought about the nature of metal ion reactivity in biological systems. The high performance of Fourier Transform Infrared Spectrometer (FT-IR) especially when coupled with an infrared data station gives to this instrument a great data manipulation capability.

Infrared spectroscopy of nucleic acids has progressed from the ability to identify a few group frequencies to the point where we can begin to consider the vibrational interactions of the entire molecule and its parts. This is made possible by comparison of spectra from a number of series of nucleic acid derivatives and their metal complexes. In addition, the spectra of the constituents, the nucleotides which consist of a base (purine or pyrimidine), a sugar (pentose) and the phosphate group are simpler. These groups react differently with the metal ions depending on the nature of the metal, the amount of the metal, the pH, the temperature and the solvent. One can examine the vibrations of the purine or pyrimidine rings, the sugar or the phosphate group almost separately. The types of vibrations which we observe include substituent group frequencies. The in-plane and out-of-plane bending modes of CH, CO and NH of the pyrimidine or purine ring vibrations which include the aromatic stretching region of the carbon-carbon and carbon-nitrogen double bonds at 1300-1650 cm^{-1}, the breathing and torsional modes at somewhat lower frequencies in the fingerprint region and the CH, NH and OH stretching vibrations at higher frequencies are given. Finally, there is the region of the carbon-carbon, carbon-nitrogen single bonds and the exocyclic and the sugar-phosphate groups in the range 400-1300 cm^{-1}. The spectral changes of the nucleic acids could also help determine the site of attack of the metal ion (1).

Even though the number of the normal vibrations in a molecule is 3N-6, where N is the number of atoms in a molecule, the nucleic acids give no more than 50 well-defined absorption bands in the mid-infrared region 400-4000 cm^{-1}. Vibrational spectroscopy is sensitive to conformational changes in biopolymers brought about by metal ions and it can be used to obtain information about the structure in both the solid and solution states. The spectra obtained with a Fourier transform infrared spectrometer are studied with differential techniques and the absorption of the H_2O can be digitally subtracted from the solution spectra. Infrared spectra of aqueous solutions can be examined now with FT-IR techniques as opposed to the old prism and grating instruments. The spectra obtained with an FT-IR are handled with differential techniques and the data are treated with a computer.

II. Nomenclature

The basic blocks of nucleic acids together with their derivatives and nomenclature are given in Table 1.

Figure 1 shows the common pairing of the bases in the nucleic acids, the Watson-Crick pairing with two hydrogen bonds in the pairs, adenine-thymine or adenine-uracil and three hydrogen bonds in the pair guanine-cytosine (2).

III. Structural Features of Nucleic Acid Constituents

The building blocks of DNA, the bases, when R=H, are linked by a sugar-phosphate bond. R is a ribose in the nucleosides and a ribose-phosphate in the case of a nucleotide. These bases are illustrated in Fig. 2.

3. Polymers of nucleotides constitute the polynucleotides linked by a phosphate diester bond from the 3'-OH of one to the 5'-OH of the next sugar. In our representation it is important to indicate clearly the 5'-end (head) and the 3'-end (tail).

TABLE 1

Common purine and pyrimidine bases and derivatives

Base Purine or Pyrimidine	Base+Sugar Nucleoside	Base+Sugar+Phosphate Nucleotide
Adenine (A)	Adenosine	Adenosine MonoPhosphate (AMP)
Guanine (G)	Guanosine	Guanosine MonoPhosphate (GMP)
Hypoxanthine (H)	Inosine	Inosine MonoPhosphate (IMP)
Cytosine (C)	Cytidine	Cytidine MonoPhosphate (CMP)
Uracil (U)	Uridine	Uridine MonoPhosphate (UMP)
Thymine (T)	Thymidine	Thymidine MonoPhosphate (TMP)

The basic structural determinant is hydrogen bonding between the base-pairs and stacking.

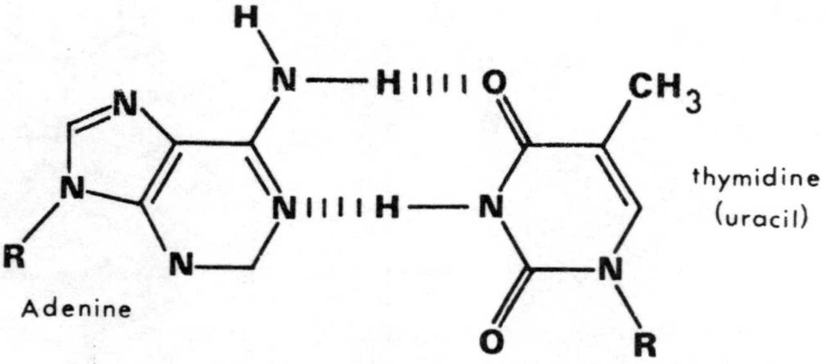

pair: A-T or A-U

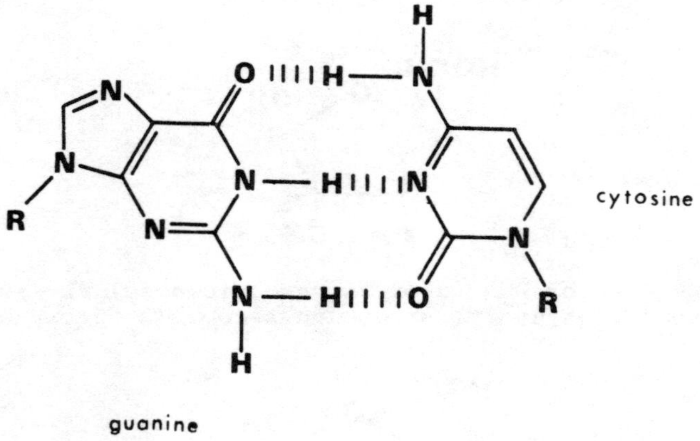

pair: G-C

Figure 1. The four base pairs used to construct the double helix hydrogen bonds (IIIII).

a) purines

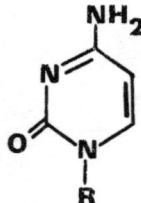
Adenine

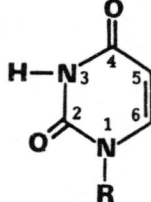
Guanine (enol form)

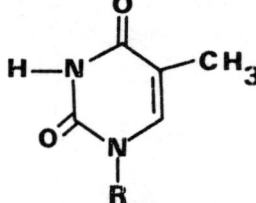
Hypoxanthine (enol form)

Uracil Cytosine (kero form) Thymine (keto form)

Figure 2. The chemical structures of the four DNA bases and derivatives

The ribose ring is represented with the numbering as follows and is found in RNA (Ribonucleic Acid).

In the case of DNA (deoxyribonucleic acid) the sugar does not contain a 2'-hydroxyl group and becomes 2'-deoxynucleoside.

Examples of a tetranucleotide represented as follows:
A-U-G-C or as pApUpGpC the stick representation:

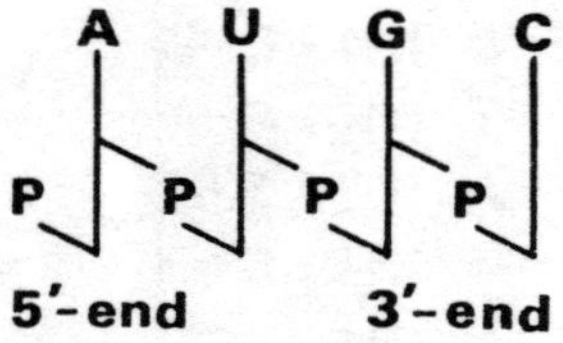

A single strand is illustrated in Figure 3.

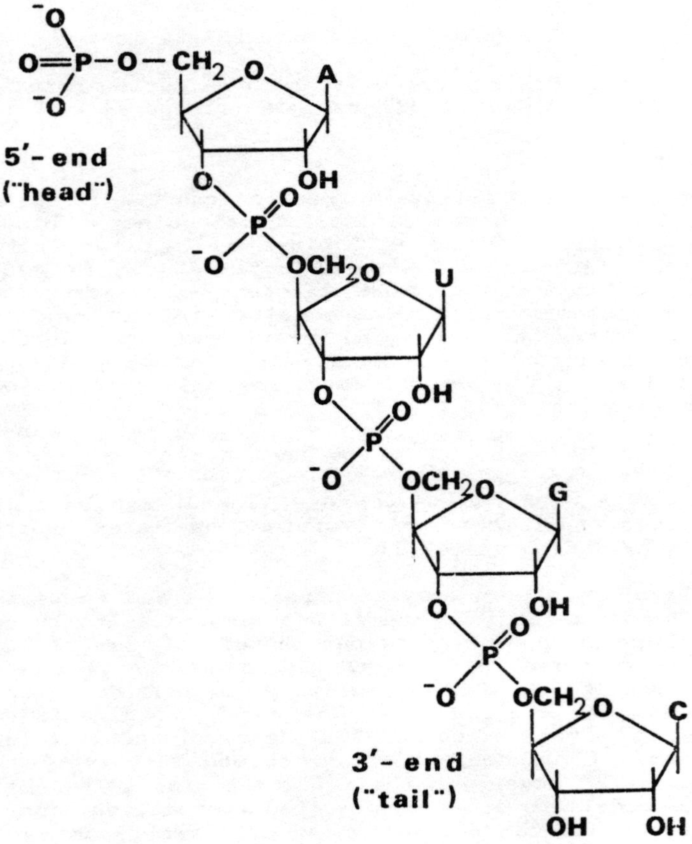

Figure 3. Single strand

A double strand or a double helix is represented as an antiparallel complementary sequence of a double stranded segment of DNA in Figure 4.

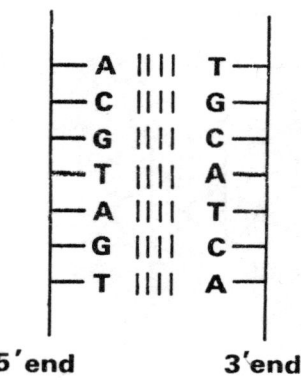

Figure 4. A schematic view of a DNA segment

Higher orders of structure exist such as nucleoproteins: For example, DNA-histones and RNA-protein (ribosomes).

IV. Metal Ions

The metal ions are potential reaction centers in biological media and react with the biological molecules by blocking basic sites. The life metal Mg^{2+} plays an important role in maintaining the integrity of several organs, such as myocardium, kidneys and bone. Magnesium deficiency may cause arteriosclerosis, thrombosis and even myocardial infarction. Many polycationic counterions are found in viruses and microorganisms, in particalar divalent metal ions strongly associate with nucleic acids. The general metal-molecule interaction is an acid-base reaction:

$$M^{n+} + L \longrightarrow M^{n+} - L$$

where, M^{n+} = metal ion, the acid; L = ligand, the base (biological molecule, nucleic acids, proteins, membranes, solvent, etc.) and n = oxidation state (3).

The role of metals in enzymatic reactions and in cells is well known. Many metal ions are strong bioactivators by coordinating to ligands, provoking oxidation-reduction reactions by being reduced or oxidized and by blocking important active sites. Some metals exhibit high activity for certain sites and can be inactive for other sites. The metal coordination complexes may also be used as models for the elucidation of fundamental problems of biochemical structure and reactivity. The intracellular metal ions with their multi-coordinating ability are beautiful models to study complicated systems, such as, the action of nerves and the alteration of cell membranes as well as the mutagenicity in nucleic acids. Inside the cell the life metals, Ca^{2+} (10^{-7}M) and Mg^{2+} (10^{-3}M) clearly form complexes with many substrates and nucleic acids. The Mg^{2+} ion is known to be bound with adenosinetriphosphate (ATP), adenosinediphosphate (ADP), nucleic acids and enzymes and its deficiency in the body during infancy and early childhood can cause manifestations of cardiovascular disorders or death. Magnesium ions

(Mg^{2+}) may be used to bind negatively charged phosphate groups in nucleic acids and thus stabilize a helix or can joint two strands together. The metals that have been found to be essential in live systems are illustrated in Figure 5. Here we see the group of metals that are found to interact and form stable complexes in living systems so far. This does not seem to be an exclusive group of metals, but their position in the periodic table is significant, because it coincides closely with the borderline region of elements between class (a) and class (b) metals and it contains the class (a) elements. It contains the elements Fe and Mo that are active sites in biological systems (4).

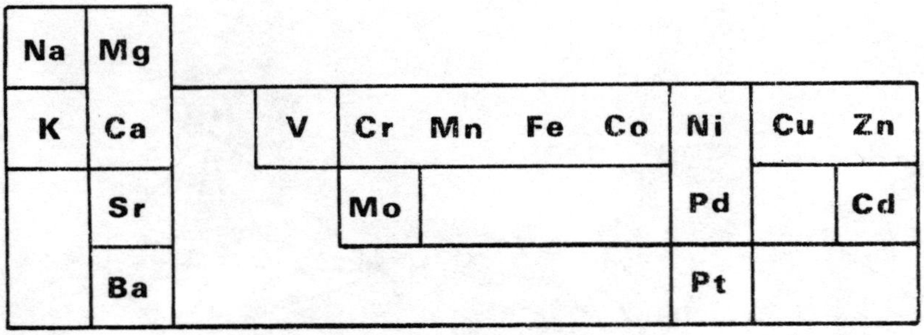

Na	Mg								
K	Ca		V	Cr	Mn	Fe	Co	Ni	Cu	Zn
	Sr			Mo				Pd		Cd
	Ba							Pt		

Figure 5. Periodic table of metals found to be essential in life systems.

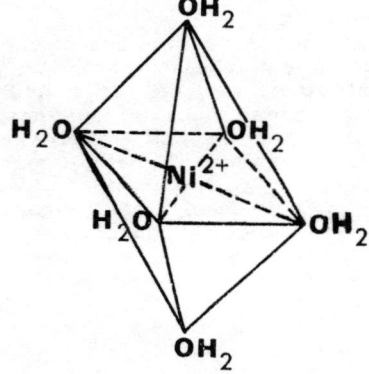

Figure 6. Geometrical configuration of a hydrated nickel metal ion with a coordination number of attached water ligands 6.

It is known that metals in a biological milieu behave as positive ions and usually they form electrovalent compounds with a tendency to form complex ions. Such ions are called coordination complexes and the ions or molecules attached to the metallic ion are called ligands. The metal ions in water solution and in the biological milieu, which is aqueous by excellence are known to form complexes with water molecules:

$$M^{n+} + 6H_2O \rightarrow M(H_2O)_6^{n+}$$

The attached molecules of water are arranged at the corners of a regular octahedron (see Fig. 6).

Complexes with a coordination number of 4 have attached ligands at the corners of a square plane or of a regular tetrahedron (see Fig. 7).

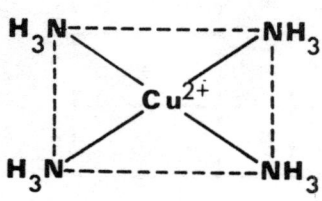 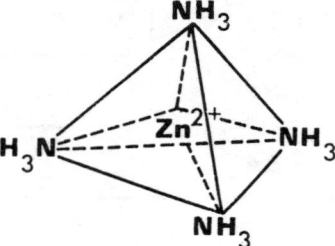

Figure 7. Geometrical arrangements with coordination number 4.

V. Sample Handling

Some of the practical difficulties in recording IR spectra of biological molecules and their metal ion complexes in aqueous solutions are reflected by the limited amount of spectroscopic data reported on these substances in water solutions. The intense absorption of water in the mid-infrared region of the spectrum restricts the regions where aqueous solutions can be studied. However, with the development of Fourier transform Infrared spectrometers the study of biological molecules and biopolymers has begun. Examples could be drawn from the extensive-biobliography on these molecules. The absorption of aqueous solutions of biological molecules is measured against water as reference.

VI. Frequencies

Only the main features of the spectra are discussed here. The hydroxyl, amine and carbon-hydrogen stretching frequencies are in the $4000-2700 cm^{-1}$ region where, one finds the NH_2, NH OH and CH vibration frequencies of the bases and the sugar. Since there is strong hydrogen bonding in both the free nucleic acids and their metal-aggregates very little information can be obtained on the nature of metal-ligand interaction in this region.

TABLE 2. FT-IR absorption bands (cm^{-1}) of free 5'-GMP with metal complexes and possible assignments.

5'GMP Na_2	Ni(GMP) $8H_2O$	Pt(NH_3)$_2$(GMP)$_2$ Cl_2	Mg(GMP) $10H_2O$	Assignment (See Table 4)
3540 sh	3550 sh	3540 m	3520 m	$\nu(NH_2)$ asym.
3512 sh	3466 s	3460 m	3480 sh	
3456 s		3450 s	3480 m	
3436 s	3410 s	3410 s	3430 s	$\nu(NH_2)$ sym.
3394 w	3371 vs	3360 s	3380 s	
3331 vs	3320 m	3340 m	3330 m	$\nu(OH)$H-bonded
3248 w	3240 m	3250 s	3240 m	
				$\nu(NH_2)$ and $2\times\delta(NH_2)$
3211 s	3210 vw	3230 m	3200 m	
	3186 s	3140 s	3160 m	
				$\nu(C_8-H)$
3145 m	3114 s	3100 s	3120 m	
3088 s	3039 vs	3070 m	3070 s	
				$\nu(NH)$
3028 s	3039 vs	3050 w	3040 m	
2987 s	2980 m	2980 w	2980 w	
				$\nu(CH_2)$ and (CH)
2981 w	2926 vw	2950 m	2960 m	
1692 bs	1690 vs	1694 vs	1697 bs	$\nu C_6=O$ $\nu C_6=C_5$
	1655 sh	1653 sh	1660 sh	
1662 sh				δNH_2 νC_2-NH_2
	1645 s	1643 s	1645 w	
1597 s	1611 s	1599 s	1611 m	$\nu C_4=C_5$ νC_4-N_3 νC_5-N_7
1575 sh	1570 m	1550 w	1560 w	$\nu C_4=C_5$ δN_1-H
1535 m	1533 m	1543 m	1536 m	νC_4-N_9 $\nu(C_6=O)$
-	1476 m	1503 m	1483 m	
1479 m	1465 sh	1460 sh	-	δC_8-H $\nu(C_8-N_7)$
1410 sh	1410 vw	1412 vw	1414 w	δCH and δCH_2
1358 s	1389 m	1383 m	1383 m	pyrimidine ring
	1352 m	1348 m	1357 m	
1333 sh				νC_8-N_9 νN_7-C_8
1254 sh	1260 w	1270	1254 m	$\nu C_8-N_7 \nu N_1-C_6 +\nu N_7-C_5$
1234 s	1236 m			νC_8-N_7 δC_8-H
	1216 sh	1218 m	1220 sh	
1204 m				νC_8-N_7 δC_8-H
	1205 s	1205 s	1206 m	
1175 sh	1180 m	1178 sh	1180 m	νC_8-N_7 νN_9-sugar
1125 sh	1109 sh	1109 vs	1107 bs	νCO of sugar ring
1070 bs	1084 vs	1084 s	1080 sh	νdeg. PO_3^{2-}
972 s	981 s	974 s	993 s	νsym. PO_3^{2-}
910 w	910 w	906 vw		ribose ring stretch
	882 w	885 w	880 w	ribose ring and
867	860 w	862 w	865 w	$ROPO_3^{2-}$ stretch
	835 vw			
805 m	803 m	798 m	803 m	P-O stretch
780 m	773 m	777 m	781 m	ribose ring stretch
682 m	705 m	734 w	728 w	breathing mode
			705 w	
	669 m	680 w	692 w	ring def.
620 m	627 m	631 m	630 w	NH out-of-plane def.
570 sh	597 m	585 m	579 w	PO_3^{2-} sym. def.
525 m	536 m	528 sh	556 m	pyrimidine ring def.
465 w	476 w	472 w	470 w	PO_3^{2-} deg. def.
430 vw	430 w	424 w	424 w	ribose ring def.
478 w	415 w			skeletal def.

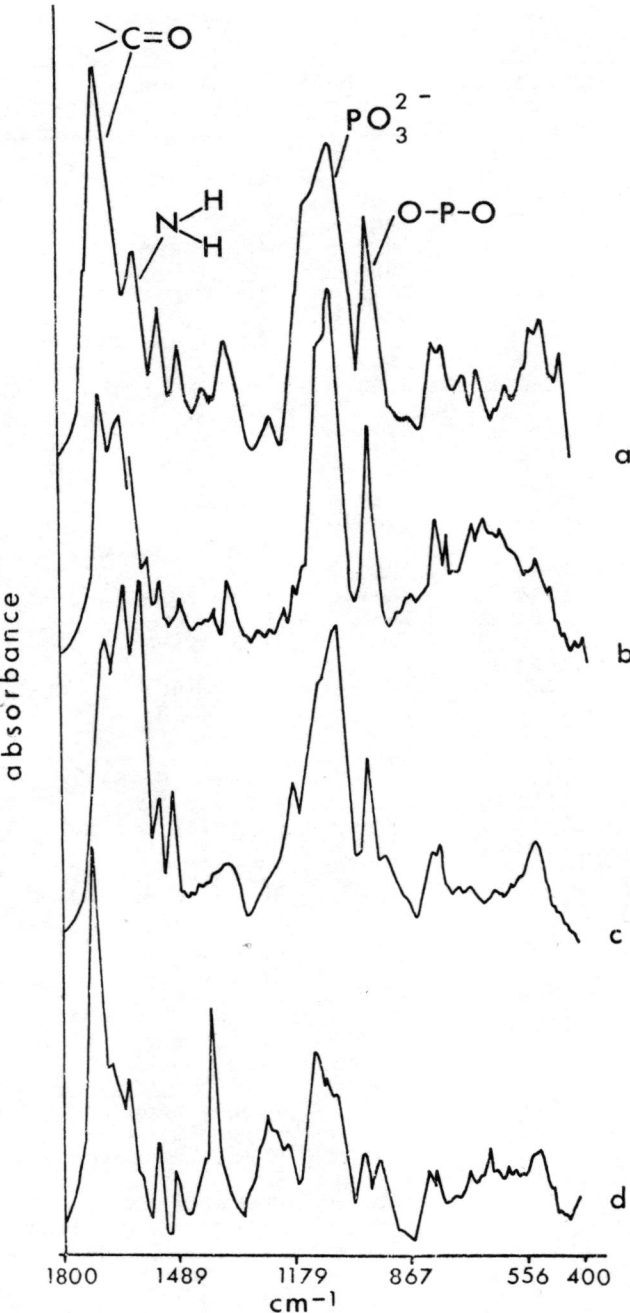

Fig. 8. FT-IR spectra of defined metal complexes of 5'-GMPNa$_2$ in the solid state, (a) 5'-GMPNa$_2$ (b) Ni(H$_2$O)$_5$(GMP) (c) Pt(NH$_3$)$_2$(GMP)$_2$Cl$_2$ and (d) Mg(H$_2$O)$_5$(GMP) 5H$_2$O.

The assignments are given in Table 2 for the NH_2, OH and CH stretching frequencies. The CH frequencies include the aromatic C_8-H, the sugar and the exocyclic $-CH_2-OPO_3^{2-}$ frequencies (5). We observe slight changes and occasional splittings in the above frequencies on metal complex formation which are characteristic of the site interaction of the metal with the biomolecule. The main features of the FT-IR spectra for example for the nucleotide 5'-GMPNa$_2$ and some metal complexes are given in Table 2 and the spectra in Figure 8. The spectral changes in that region are considerable upon complex formation with the metal bound at a specific site of the nucleotide(1).

VII. Effect of Metal Ions on Frequencies and Intensities

FT-IR spectroscopy is complementary to Raman spectroscopy and the use of the Raman technique together with the infrared should always be considered. The IR spectra of DNA have been recorded and analysed (6). FT-IR spectra of DNA and the effect of a heavy metal, cis-Pt(NH$_3$)$_2$Cl$_2$ on specific vibrations is shown in Figure 10. Important changes occur in the regions and 1700-1500 and 1200-900 cm^{-1} in the shape of the carbonyl and amide band contour, as well as, in the phosphate or ribose-phosphate bonds and in particular the intensity of the vibration at 1054 cm^{-1} (7) (Table 3, Fig. 9) diminishes drastically even with .03 Pt/nucleotide.

TABLE 3

Major Raman and IR frequencies (cm^{-1}) in B-DNA and B-DNA-cis-platinum of the ribose-phosphate vibrations.

Observed in B-DNA Raman (10) solution	IR B-DNA (7) film	B-DNA + cis platinum(7) film	Calcd.(11)	Assignment[a]
	1200 vs	1220 vs	1209	antisym. PO$_2^-$ stretch
			1200	ribose-phosphate
1140			1160	ribose-phosphate
			1112	C-O stretch
1094	1088 ms	1085 s	1094	sym. PO$_2^-$ stretch
1051	1054 ms	1060 m	1058	ribose phosphate
1015	1018 w	1022 w	999	ribose
920	970 m	972 m	959	C-C stretch
879	880		886	ribose-phosphate
835	835		834	antisym. O-P-O str.
786	786		789	sym. O-P-O stretch

The Raman vibration of the symmetric stretching of the phosphodiester group -5'-CH$_2$-O-PO$_2$-O- appears at 814 cm^{-1} and is slightly altered on fixation of a metal to the phosphate group. It has been shown (8) that the intensity of the Raman line at 814 cm^{-1} is a measure of the conformation in the tRNA backbone. The intensity of the above line undergoes a decrease of about 20% when the Mg^{2+} ions are extracted with a chelating agent (EDTA) and suitable dialysis. The backbone structure is

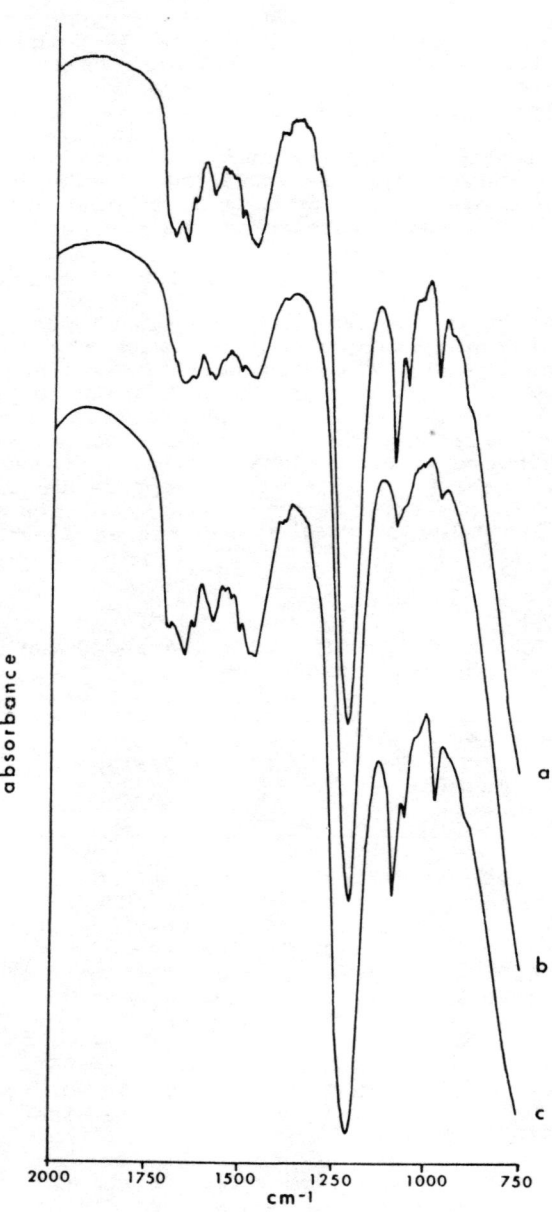

Fig. 9. The FT-IR spectra of calf thymus DNA and its cis-platinum complexes (a) control (b) with .20 and (c) with .05 Pt/nucleotide.

disordered by changes in the temperature or the ionic environment, the line at 814 cm^{-1} changes both in intensity and position and a marked decrease in intensity is observed at 814 cm^{-1}. The role of the metal in general is to melt the double helix and to decrease the disorder in the backbone. It is known, for example, that the intensities of the Raman bands at 690, 725 and 785 cm^{-1} are related to the stacking of the nucleic acid bases G, A and C or U, respectively. This conclusion was reached also from UV-VIS spectra, where an hypochromism was observed on stacking. It is found that the line at 690 cm^{-1} of G, decreases when the Mg^{2+} ions are removed from tRNA. The decrease is of the order of (40%) which implies that about eight or nine of the 22 G residues in the tRNA are unstacked by removing the Mg^{2+} ions. The line at 725 cm^{-1} due to A, however shows an increase in the stacking of adenine by removing Mg^{2+}. The conclusion that magnesium ions are stabilizing specific conformations in nucleotides was reached also from high resolution proton NMR studies (9) which are in agreement with the vibrational studies.

VIII. Phosphate Frequencies

For nucleic acids the phosphate frequencies of the PO$_2^-$ group in the backbone chain of DNA, common to the sugar-phosphate linkage, occur at 1200, 1090 cm^{-1}, 970 and 790 cm^{-1} (see Table 3).

These vibrations are conformation dependent and may be perturbed by metal ions. The vibrations of the group, PO$_2^-$ are given in Table 3. The anionic oxygens in the (OPO)$^-$ plane can interact with metal ions, in particular, with hard metals of class (a), for instance, the alkaline metals, Li$^+$, Na$^+$, K$^+$, or the alkaline-earth metals, Mg^{2+}, Ca^{2+}, Ba^{2+} or the cationic species in general, i.e., spermine (NH$_3^+$CH$_2$CH$_2$NH$_2^+$CH$_2$CH$_2$NH$_2^-$CH$_2$CH$_2$NH$_3^+$) in order to stabilize a particular conformation of the base sequence. The association of these cations with nucleic acids may affect rotamer populations of the possible rotamer equilibria that might exist in solution, such as rotation around the ω or the φ bonds in the case of nucleotides (Fig. 11) (10).

The phosphate absorption bands may be clearly identified by investigation of the fixation of these cations to the phosphate chain. Changes that are brought about by complex formation can help assign these absorptions. The degenerate stretching band, of PO$_3^{2-}$ in nucleotides which absorbs at 1110-1000 cm^{-1} is split by metal-phosphate-complex formation (5). In solution the extent of the splitting may be an indication of the percentage of the conformational species existing in solution by coordination of the metal ion to the PO$_3^{2-}$ group. The antisymmetric stretching absorption of the

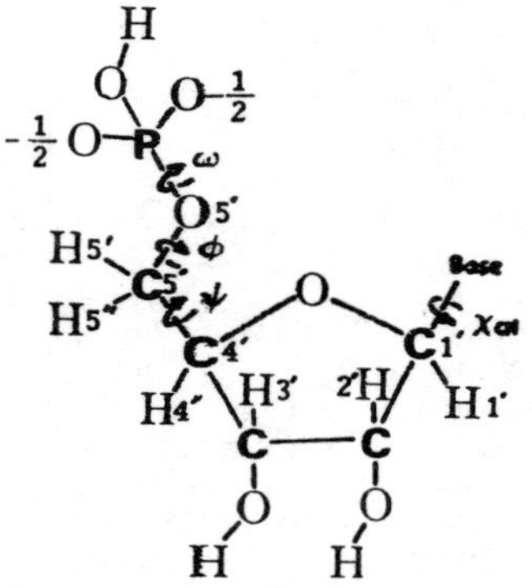

Fig. 10

PO_2^- group at 1200 cm^{-1} in polynucleotides or DNA is also perturbed upon metal interaction. This perturbation may indicate some coordinative interaction between the metal ion and the PO_2^- group or that this group has been rotated in such a way as to bring closer to the positive metal ion its anionic phosphate oxygens. The FT-IR spectra of calf thymus DNA sodium salts and some metal complexes are shown in Fig. 11. The spectra shows considerable intensity changes of the phosphate group frequencies upon metal interaction (see Table 3 and Fig. 11).

On the basis of FT-IR studies on metal-nucleic acids it can be concluded that fixation of a metal ion to a particular site of a nucleotide causes drastic spectral changes of the vibrations associated with that site. The complex vibrational modes of these systems are coupled and assignments are diffi-

TABLE 4

Major Raman and IR frequencies (cm^{-1}) in 5'-GMPNa$_2$ and $CH_3OPO_3Na_2$ of the phosphate vibrations

Observed in 5'-GMPNa$_2$		Assignments	Observed in $CH_3PO_3Na_2$ (14)		Calcd
Raman(12,13) solution	IR (5) solid		Raman solution	IR solid	
1088 m	1070 bs	PO_3^{2-} deg stretch.	–	1115	1097
1046 w	–	C-O stretch.	1057 d	1056	1056
980 w	972 s	PO_3^{2-} sym. stretch.	985 p	983	992
810 m	802 m	P-O$_2$ sym. stretch.	750 p	750	751
–	–	PO_3^{2-} sym. def.	–	573	596
–	540	PO_3^{2-} rock and PO_3	–	510	508
–	460 sh	"deg" def.	–	393	396

[a] asym. asymmetric, sym.: symmetric, def.: deformation, deg.: degenerate, d: depolarized, p: polarized, vs: very strong, s: strong, b: broad; ms: medium-strong, m: medium, w: weak, sh: sharp.

cult. However, metal complexation often can help in the assignment of the bands, that are related to the site of reactivity.

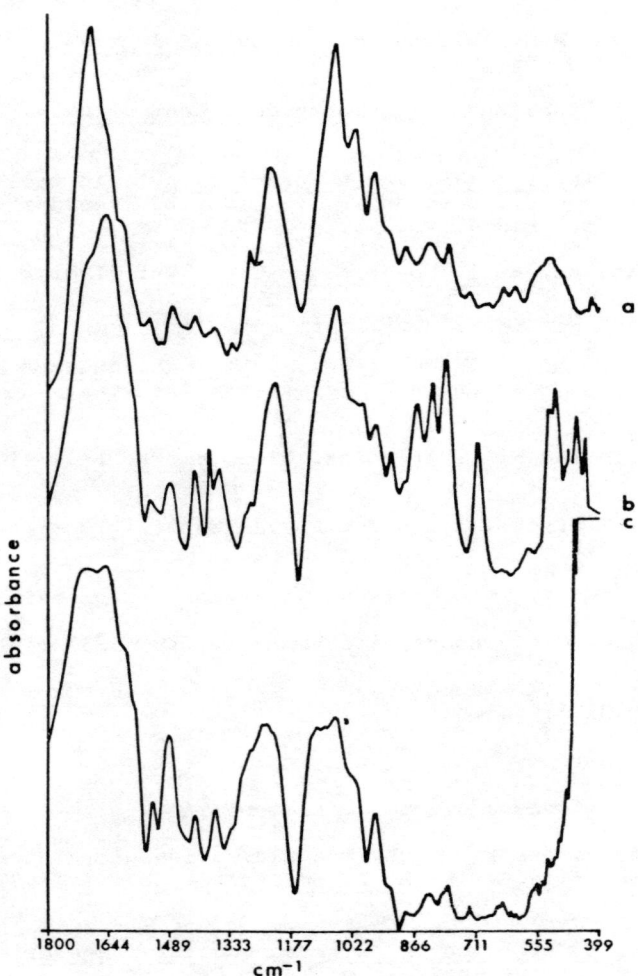

Fig. 11. FT-IR spectra of calf thymus DNA (a) and its metal complexes in the solid state (b) copper and (c) magnesium.

References

1. T. Theophanides (Ed.) "Infrared and Raman Spectroscopy of Biological Molecules", 1979 by D. Reidel Publishing Company, Dordrecht, Holland; idem, Can. J. Spectrosc., 26, 165 (1981).

2. J.N. Davidson, "The Biochemistry of the Nucleic Acids", 1972 by Chapman and Hall and Science Paperbacks.

3. G.L. Eichhorn, "Inorganic Biochemistry, Chaps. 33 and 34, Elsevier, 1975.

4. F. Basolo and R.G. Pearson, Mechanisms of Inorganic Reactions, 2nd ed., Wiley, 1976.

5. H.A. Tajmir-Riahi and T. Theophanides, Can. J. Chem. In press, 1983.

6. M. Tsuboi, "Infrared and Raman Spectroscopy", in Basic Principles in Nucleic Acids, P.O.P. Tso, Ed. (Academic Press, New York, 1974, Vol. 1, pp. 399-452.

7. T. Theophanides, Appl. Spectroscopy, 35, 461 (1981).

8. R.C. Lord, Appl. Spectroscopy, 31, 187 (1977).

9. M. Polissiou and T. Theophanides, in "Biomolecular Stereodynamics", Vol. II, Ed. R.H. Sarma, Adenine Press, New York, p. 487.

10. S.C. Erfurth and W.L. Peticolas, Biopolymers, 14, 347 (1975).

11. R.C. Lu, E.W. Prohofsky, and L.L. Van Zandt, Biopolymers, 16, 2491 (1977).

12. M. Manfait and T. Theophanides, Magnesium, In press, 1983.

13. M.J. Lane and G.J. Thomas, Biochemistry, 18, 3839 (1979).

14. T. Shimanouchi, M. Tsuboi, and Y. Kyogoku, Adv. Chem. Phys., 7, 435 (1964).

Additional Bibliography

A. Sampling techniques including microsampling

1. P.R. Griffiths and F. Block, "Considerations for Infrared Ultramicrosampling". F. Appl. Spectrosc., 27, 431-435 (1973).

2. S.T. King, "Application of Infrared Fourier Transform Spectroscopy to the Analysis of Micro Samples".

3. D.H. Anderson and T.E. Wilson, "Novel Approach to Micro Infrared Sample Preparation". Anal. Chem., 47, 2482-2483 (1975), IR-21.

4. T. Hirschfeld and K. Kizer, "Direct Recording or Intermolecular Interaction Perturbation Spectra by Fourier Transform Spectroscopy". Appl. Spectrosc., 29, 345-351 (1975).

5. T. Hirschfeld, "Optimum Diameter of Fourier Transform Infared Microsampling Cell". Appl. Spectrosc., 30, 353-354 (1976).

6. T. Hirschfeld, "Subsurface Layer Studies by Attenuated Total Reflection". Fourier Transform Spectroscopy (1976).

7. R.P. Oertel and A.J. Fehl, "Combination Glove Box-Sample Compartment Cover for A fourier Transform IR Spectrophotometer". Rev. Sci. Instrum., 46, 855-856 (1975).

8. C.T. Reddy and A.C. Gilby, "An Infrared Beam Projector For Use in Fourier Transform Spectroscopy". Appl. Spectrosc., 29, 532 (1975).

9. W.D. Stephens, W.W. Schwarz, R.B. Kruse, D.F. Vandiver, A. Mantz, K.L. Kizer, "Application of Fourier Transform Spectroscopy to Propellant Service Life Prediction". AAIA/SAE 12th Propulsion Conference, Palo Alto, CA (July 26-29, 1976) IR-22.

10. M.P. Fuller and P. Griffiths, "Diffuse Reflectance Measurements by Infrared Fourier Transform Spectroscopy". Anal. Chem. (Nov. 1978) IR-27.

11. N. Wright and W.C. Lee, Nature., 136, 300 (1935).

12. J.T. Edsall, J. Phys. Chem., 41, 135 (1937).

B. General References on Biological Molecules (Infrared and Raman)

13. T. Theophanides, "Infrared and Raman Spectroscopy of Biological Molecules, D. Reidel Dordrecht: Holland, 1979, 187, 205.

14. R.C. Lord and G.J. Thomas, Dev. Appl. Spectrosc. 6 (1968).

15. B. Fanconi, B. Tomlinson, L. Nafie, W. Small and W. Peticolas, J. Chem. Phys., 51, 3993 (1969).

16. P. Griffiths, Chemical Infrared Fourier Transform Spectrosc., Wiley, New York, NY, 1976.

17. N.T. Yu, Critical Rev. in Biochem., 229 (1977).

18. R.C. Lord and N.T. Yu, J. Mol. Biol., 50, 509 (1970).

19. R.C. Lord and N.T. Yu J. Mol. Biol., 51, 203 (1970).

20. R.C. Lord, Pure Appl. Chem., 28 (1971).

21. R.C. Lord, A.M. Bellocq and R. Mendelsohn, Biochim. Biophys. Acta, 257, 280 (1972).

22. R.C. Lord, M.C. Chen and R. Mendelsohn, Biochim. Biophys. Acta, 328, 252 (1973).

23. J.L. Koenig and B.G. Furshour, Biopolymers, 11, 2505 (1972).

24. N.T. Yu, C.S. Culver and D.C. O'Shea, Biochim. Biophys. Acta, 263, 1 (1972).

25. N.T. Yu and C.S. Liu, J. Am. Chem. Soc., 94, 3250 (1972).

26. B.G. Frushour and J.L. Koenig, Advances in IR and Raman Spect. Vol. I, Heyden, New York, NY, 1975.

27. G.J. Thomas, Vibrational Spectra and Structure Vol. III, Marcel Dekker, ed. by J.R. Durig, New York, NY, 1974.

28. J.L. Lippert, D. Tyminski and P.J. Desmeules, J. Am. Chem. Soc., 98, 7075 (1976).

29. N. Greenfield and G.D. Fasman, Biochemistry, 8, 4108 (1969).

30. J.G. Guillot, M. Pezolet and D. Pallota, Biochim. Biophys. Acta, 491, 423 (1977)

31. N.T. Yu, E.J. East and R.C.C. Chang, Exp. Eye Res., 24, 321 (1977).

32. R.K. Sharma, R.L. Kisliuk, S.P. Verma and D.F.H. Wallach, Biochim. Biophys. Acta, 391, 19 (1975).

33. W.L. Peticolas, Biochimie, 57, 417 (1975).

34. T.A. Turano, K.A. Hartman and G.J. Thomas, J. Phys. Chem., 80, 1157 (1976) and papers cited within.

35. G.J. Thomas, Appl. Spectrosc., 30, 483 (1976).

36. M. Shie, E.N. Dobrov, T.I. Tikchonenko, Biochem. Biophys. Res. Communic., 81, 907 (1978).

37. L.D. Esposito and J.L. Koenig, Fourier Transform IR Spectroscopy Vol. I, Acad. Press, NY 1978.

38. T.G. Spiro and T.C. Strekas, J. Am. Chem. Soc., 96, 338 (1974) and references cited within.

39. T.G. Spiro, Acc. Chem. Res., 7, 339 (1974).

40. T.G. Spiro, Vibrational Spectra and Structure Vol. V, Elsevier, ed. by J.R. Durig, Amsterdam and New York, 1976.

41. R.C. Lord and G.J. Thomas, Spectrochim. Acta, 23A, 2551 (1967).

42. R.C. Lord and G.J. Thomas, Biochim. Biophys. Acta, 142, 1 (1967).

43. W.H. Woodruff and G.H. Atkinson, Anal. Chem., 48, 186 (1976).

44. P.R. Carey and H. Schneider, Acc. Chem. Res., 11, 122 (1978).

45. T. Theophanides, Can. J. Spectrosc., 26, 165 (1981).

C. Aqueous Solutions and Biological Systems

46. A.W. Mantz and H.K. Morita, "Fourier Transform Infrared Spectroscopic Evidence for Soil Linuron Interation, Appl. Spectrosc., 30, 587 (1976).

47. R.J. Jakobsen and M.R. Gendreau, "Blood Plasma/Implant Interfaces. FT-IR Studies of Adsorption on Polyethylene and Heparin-Treated Polyethylene Surfaces". Artificial Organs, 2, 183 (May 1978).

48. R.M. Gendreau, R.J. Jakobsen, "Fourier Transform Infrared Techniques for Studying Complex Biological Systems". Appl. Spectrosc., 32, 326 (1978).

49. J.O. Alben, S.S. Choi, A.D. Adler, W.S. Caughey, "IR Spectroscopy of Porphyrins". Annals of the New York Academy of Sciences, 206, 278 (Oct. 1973).

50. J.O. Alben and L.Y. Fager, "Structure of the Carbon Monoxide Binding Site of Hemocyanins Studied by Fourier Transform Infrared Spectroscopy". Biochemistry, 11, 4786 (1972).

51. J.O. Alben, J.A. Berzofsky, P. Pelsach, "Surfheme Proteins III. Carboxysulfmyoglobin: The Relation Between Electron Withdrawal from Iron and Ligand Binding". J. of Bio. Chem., 247, 3774 (1972).

52. P.S. Callahan, "Studies on the Shootborer". Turrialba, 23, 263 (1973).

53. J.L. Koenig and D.L. Tabb, "Infrared Spectra of Globular Proteins in Aqueous Solution". (1974).

54. M.J.D. Low and R.T. Yang, "Ir Spectra of H_2O in Aqueous Solutions Using Internal Reflection Spectroscopy". Spectroscopy Letters, 6, 299 (1973).

55. M.J.D. Low and R.T. Yang, "The Measurement of Infrared Spectra of Aqueous Solutions Using Fourier Transform Spectroscopy". Spectrochimica Acta, 29A, 1761 (1973).

56. M.J.D. Low and R.T. Yang, "Quantitative Analysis of Aqueous Nitrite/Nitrite Solutions by Infrared Internal Reflectance Spectrometry". Anal. Chem., 45, 2014 (1973).

57. R.T. Yant and M.J.D. Low, "Infrared Internal Reflection Spectra of Methanol-Water Mixtures". Spectrochimica Acta, 30a, 1787 (1974).

58. R.P. Oertel, H.C. Smitherman and A.J. Fehl, "Fourier Transform Infrared Determination of the Rate of Oxygen Exchange between Acetone and D_2O". Appl. Spectrosc., 29, 195 (1975).

59. P.S. Callahan, "Insect Antennae with Special Reference to the Mechanism of Scent Detection and the Evolution of the Sensilla", <u>Int. J. Insect Morphol & Embryol</u>, <u>4</u>, 381 (1975).

60. M. Gomez-Taylor, D. Kuehl and P.R. Griffiths, "Vibrational Spectrometry of Pesticides and RElated materials on Thin Layer Chromatography Adsorbents". <u>Appl. Spectrosc.</u>, <u>30</u>, 447 (1976).

THE APPLICATION OF FOURIER TRANSFORM INFRARED SPECTROSCOPY
TO THE STUDY OF MEMBRANES.

Henry H. MANTSCH

National Research Council Canada, Division of Chemistry,
Ottawa, K1A 0R6, Canada.

INTRODUCTION

Infrared spectroscopy has been a late addition to the spectroscopic inventory of the membrane biophysicist. The reason has been the presence of water. Biological membranes not only encompass a range of widely different molecular structures, but like most biological structures require an aqueous environment whereby water is not only the solvent of choice, but often part of the molecular structure itself. Water which does not impair spectroscopic measurements using the NMR, ESR, UV or Raman techniques is a strong infrared absorber, a fact which has precluded or severely limited the application of conventional infrared spectroscopy to the study of biological systems.

However, the field of infrared instrumentation has changed very much over the last few years. Fourier-transform infrared (FT-IR) spectrometers are now widely available, including sophisticated low-cost models which are quite suitable for the study of biological preparations. The advantages of FT instruments over conventional dispersive instruments are well documented and need not be discussed here (1-4).

MEMBRANE STRUCTURE

Membranes are the most common cellular structures; they are recognized as being involved in almost all aspects of cellular activity ranging from simple functions such as food entrapment in unicellular organisms, to very complex functions such as immunorecognition in higher organisms. This functional diversity rests on a structural diversity, which in turn is reflected in the wide variety of lipids and proteins that compose different membranes. Thus, an understanding of the physical principles that govern the molecular organization of membranes is essential for an understanding of their physiological roles.

The present commonly accepted view of a biological membrane is based on the fluid mosaic model of Singer and Nicolson which envisages membranes as

composed of proteins incorporated either wholly or partly in a fluid-like sea
of bilayer lipids; the latter provide the structural framework of the membrane
and allow the proteins to move about more or less freely in the lipid matrix.
An excellent introduction to the subject of cell membranes can be found in the
book edited by Weissmann and Claiborne (5).

TEMPERATURE DEPENDENT STRUCTURAL CHANGES:
THE THERMOTROPIC MESOMORPHISM OF MEMBRANES

Physical studies of biomembranes have concentrated mainly on the
organization of the lipid matrix. Perhaps the best studied physical property
of membranes is their thermotropic mesomorphism. This change of state induced
by temperature is reflected in an order-disorder phase transition of the lipid
matrix and is commonly referred to as the gel to liquid crystal phase
transition.

For the study of the thermotropic mesomorphism of membranes the sample
preparation is very important. Lipid dispersions are generally prepared by
mixing the desired amounts of solid lipid and solvent (H_2O, D_2O or buffer
solution). When using D_2O as solvent a closed vessel should be used in order
to minimize exchange with atmospheric water. In order to ensure proper
hydration the lipid water mixture should be first heated above the temperature
of the gel to liquid crystal phase transition of the corresponding lipid,
then cooled before commencing the measurements. The study of natural membranes
involves re-hydration only if the membranes are isolated from the other cell
components and lyophilized (6-8). The experiments on bacterial membranes
currently under way in our laboratory utilize live bacteria or isolated
membranes which have not been lyophilized, therefore no hydration procedure is
involved (9).

Infrared spectroscopic experiments directed towards studying the
thermotropic phase behavior of membrane lipids involve collecting the spectrum
of the same system at various temperatures and monitoring changes in band
parameters as a function of temperature. This process can be brought
completely under the control of the spectrometer computer, which records a
spectrum, increments the temperture, waits for temperature equilibration then
records another spectrum (10,11).

Since the output of modern spectrometers is digitized the task of data
processing is greatly simplified. Frequency and bandwidth values can now be
determined routinely with uncertainties of less than ± 0.05 cm^{-1} (12). Among
the various data reduction methods, the technique of band narrowing or
deconvolution (whereby a known lineshape is removed from the spectra), is

extremely useful for the study of band contours comprised of more than one component; the deconvolved spectrum has narrower bands but the correct integrated intensities and frequencies are retained (13-15).

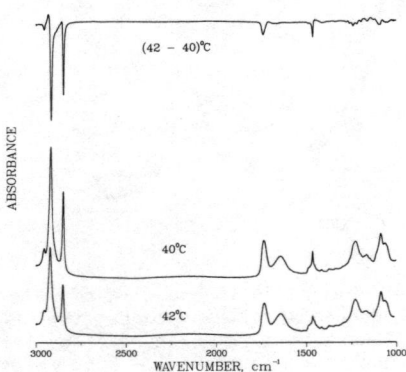

Figure 1. Infrared spectra of dipalmitoyl lecithin at 40 and 42°C (bottom) and corresponding difference spectrum (top, absorbance scale x 2). Spectra were recorded in 6 μm thick BaF_2 cells using a Digilab FTS-11 Fourier transform spectrometer equipped with a mercury cadmium telluride detector.

The temperature-induced structural changes which occur in lipid membranes can be studied by monitoring various infrared absorption bands as a function of temperature. The gel to liquid crystal phase transition involves a large structural rearrangement of the lipid bilayer and produces considerable changes in the infrared spectrum. Figure 1 shows the infrared spectra of lecithin membranes at 40 and 42°C along with the corresponding difference spectrum. It is quite evident, particularly from the difference spectrum, that the phase transition at 41°C induces changes in many of the infrared bands.

Vibrational modes of major diagnostic value are the CH_2 stretching bands in the 2800-3100 cm^{-1} region, the C=O stretching bands in the 1700-1800 cm^{-1} region and the CH_2 bending bands around 1470 cm^{-1}. Since these bands originate from vibrations of different molecular moieties it is possible to refer separately to "head group" and "hydrocarbon" spectral regions which probe the hydrophilic, respectively the hydrophobic parts of the lipid assemblies. The changes observed in all of these vibrational modes have been used extensively to characterize the nature of the various structural rearrangements occurring in model and natural membranes (16-26).

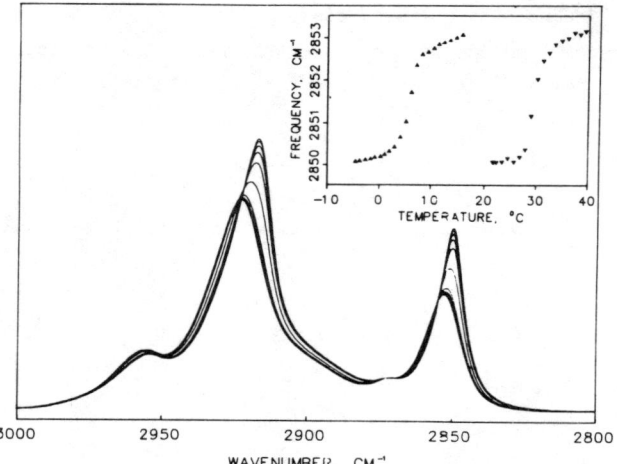

Figure 2. Temperature-induced changes in the spectra of palmitoyl lysolecithin in the region of the carbon-hydrogen stretching bands; the peak height of the infrared bands decreases with increasing temperature. The inset displays the temperature-frequency relationship of the 2850 cm^{-1} band in palmitoyl (left side curve) and stearoyl lysolecithin (right side curve).

A representative example is shown in Figure 2 which displays a series of infrared spectra of egg yolk lysolecithin at different temperatures. Lysolecithins are natural intermediaries in lipid metabolism and exist in membranes during phospholipid turnover. The strong bands at 2920 and 2850 cm^{-1} which are common to compounds with long methylene chains are the antisymmetric and symmetric CH_2 stretching modes, respectively; the weaker bands at 2955 and 2872 cm^{-1} are the asymmetric and symmetric CH_3 stretching modes of the terminal methyl group. All of these bands exhibit temperature-dependent variations in frequency and width which can be related to structural changes at the molecular level. Illustrated in the inset in Figure 2 is the temperature dependence of the frequency of the 2850 cm^{-1} band in palmitoyl and stearoyl lysolecithin. The midpoint of such frequency vs. temperature curves has been shown to correspond to the chain melting phase transition temperatures of phospholipids. The transition temperatures for palmitoyl and stearoyl lysolecithin are 5 and 28°C, respectively (27).

Figure 3 gives the detailed temperature profile for the frequency of the 2920 cm^{-1} band in palmitoyl lysolecithin. The frequency of this vibrational mode changes slightly with temperature below 0°C and again above 8°C, but shows a steep increase between 0 and 8°C. The values at temperatures below 0°C (~2917 cm^{-1}) are characteristic of conformationally highly ordered acyl chains as found in solid hydrocarbons, whereas the values at temperatures above 8°C (~2923 cm^{-1}) are characteristic of conformationally disordered acyl chains with a high content of gauche conformers such as those found in liquid hydrocarbons.

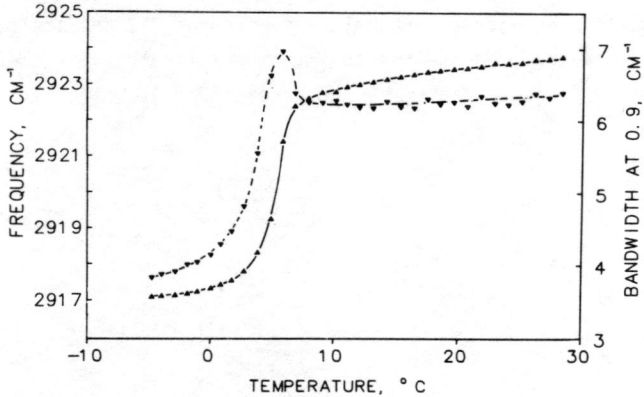

Figure 3. Temperature profiles of the frequency (solid curve) and width (broken curve) of the antisymmetric CH$_2$ stretching band of palmitoyl lysolecithin. The frequencies were obtained by determining the centre of gravity of a 2 cm^{-1} wide segment of the bands (12). Bandwiths were measured close to the top of the bands, i.e. at 9/10 peak height.

There is ample evidence in the literature that the frequencies of the CH$_2$ stretching bands of acyl chains depend on the degree of conformational disorder and hence can and have been used to monitor the average trans/gauche ratio in such systems (20-27); the higher this frequency the higher the conformational disorder.

The width of the CH$_2$ stretching band increases drastically throughout the temperature range of the phase transition, but is almost constant below and above the transition temperature. The large increase in bandwidth is typical for the melting of solid hydrocarbons and for the acyl chain melting phase transition of aqeuous phospholipids. However, from Figure 3 it can be seen that the changes in frequency and width of the CH$_2$ stretching band are not concerted. The maximum rate of change of the bandwidth is observed at a lower temperature than that of the frequency. Furthermore, a decrease in bandwidth is coupled with an increase in frequency at the high-temperature end of the phase transition i.e. between 6 and 8°C. This behavior can be explained through the overlap of bands from two species with very different bandwidths. We have previously observed this phenomenon in aqueous surfactants (28-30); it is indicative of a gradual change from the solid-like structure with narrow bands, to a micellar phase with much broader CH$_2$ stretching bands.

In certain phospholipid membranes (which contain unsaturated acyl chains) obtained from egg yolk or from human erythrocytes, the typical lamellar liquid crystalline phase converts upon further heating to a micellar non-lamellar phase. It has been shown that the driving force behind this thermally induced transition requiring a major structural rearrangement, is the increasing concentration of gauche bonds in the acyl chains which triggers the formation of the non-bilayer phase at higher temperatures (20,21). Since this phase

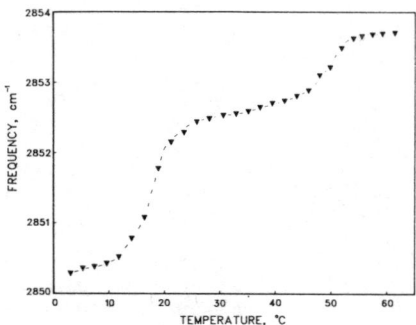

Fig. 4. Temperature-dependence of the frequency (measured as centre of gravity) of the symmetric CH_2 stretching bands in the spectra of lipid membranes obtained from egg yolk phosphatidylethanolamine.

transition involves a further increase in the gauche population of the liquid-crystalline phase, the frequency of the CH_2 stretching mode is an adequate monitor of such a transition. The plot for egg yolk phosphatidylethanolamine in Figure 4 shows a frequency shift of about 2 cm^{-1} at 18°C associated with the gel to liquid crystal phase transition and a further frequency shift of about 1 cm^{-1} at 50°C associated with the transition to the micellar phase. Both transitions are reversible.

TIME DEPENDENT STRUCTURAL CHANGES:
MEMBRANE METASTABILITY.

When hydrated lecithin membranes are annealed at temperatures near 4°C a transformation of the gel phase occurs. Figure 5 shows the changes observed in the infrared spectra of the gel phase of sulfocholine membranes when kept at 5°C for six days. Major changes are evident in head group modes such as the C=O stretching bands between 1700-1800 cm^{-1} and in acyl chain modes such as the CH_2 scissoring bands around 1470 cm^{-1} and the CH_2 wagging band progression in the 1200-1350 cmd^{-1} region. Upon incubation one observes a general narrowing and increase in peak height of all bands associated with acyl chain vibrations which is indicative of a reduction of the mobility of acyl chains (31).

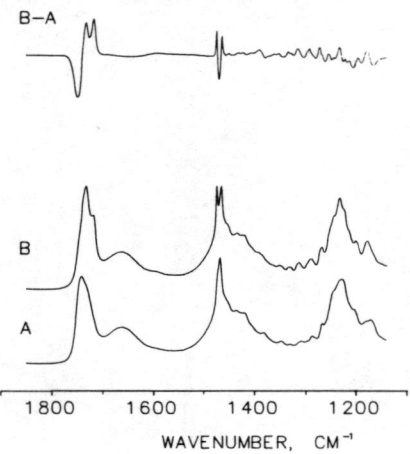

Figure 5. Infrared spectra of aqueous dipalmitoyl phosphatidylsulfocholine membranes recorded six hours after the sample preparation (trace A) and after being kept at 5°C for six days (trace B). Superimposed is the difference spectrum (B-A) which reflects the changes resulting from the incubation.

The metastability of the gel phase can be monitored by following the structural changes of the C=O stretching band contour at constant temperature as a function of time. Figure 6A shows in detail the changes induced in this band upon incubation at 5°C. The peak at 1743 cm^{-1} decreases and those at 1732 and 1715 cm^{-1} increase with longer incubation periods. Furthermore, an isosbestic point is observed at 1738 cm^{-1}.

Figure 6B shows the same spectra following Fourier self-deconvolution with a Lorentzian line of 8.5 cm^{-1} halfwidth which results in a reduction of the bandwidth by a factor of 2.2 (for details of this deconvolution procedure see refs. 13-15). From the spectra after self-deconvolution it can be seen more clearly that the C=O stretching band contour changes progressively with time and that after six days of incubation it is comprised of four narrow bands at 1744, 1737, 1730 and 1715 cm^{-1}. In fact, the infrared spectra of annealed sample resemble those of a crystalline non-hydrated sample.

The effects of this transformation can be reversed by increasing the temperature through a phase transition, known as the subtransition (32). The infrared spectra show that after incubation the acyl chain packing has changed resulting in a more rigidly packed gel phase and that the hydration shell of the head group is greatly reduced, suggesting that the forces causing the transformation are primarily changes in hydration (23,31,33).

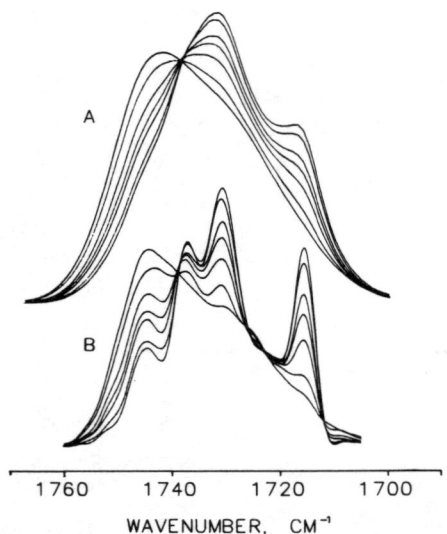

Figure 6. Time-dependent changes of the infrared spectra of dipalmitoyl phosphatidylsulfocholine membranes in the C=O stretching region. All spectra were recorded at 5°C after 0.25, 1,2,3,4,5 and 6 days at 5°C. The differences between spectra A and B are explained in the text.

DEUTERATED BIOMEMBRANES

Previous infrared spectroscopic studies had concentrated on the elucidation of the thermotropic properties of lipids in model membranes. The experience gained with the models in now actively under application to intact biological membranes. One way to circumvent the complexity of natural membranes is to use lipids with deuterated acyl chains (34-36).

Early success in the study of natural membranes was attained by the use of the relatively simple organism Acholeplasma laidlawii (6-8). It has a single membrane, the plasma membrane, and will accept deuterium labelled fatty acids from a growth medium without significant alterations. For some cases enrichment of the membrane lipids to greater than 90% deuterium containing fatty acids can be achieved by use of the protein avidin in the growth medium (8).

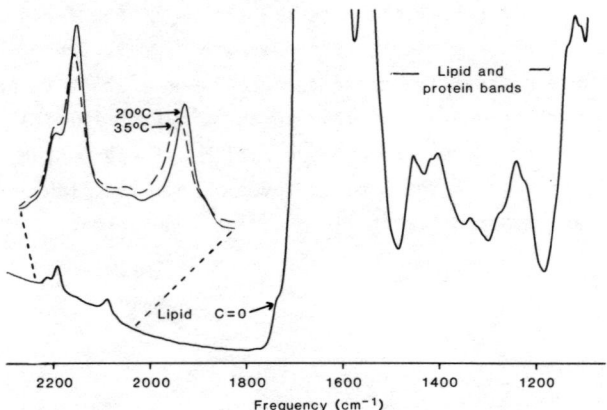

Figure 7. Infrared spectrum of live bacteria (Acholeplasma laidlawii) grown at 30°C on a diet of perdeuteriomyristic acid. The inset shows, at an enlarged scale, the C-D stretching region at 20 and 35°C.

Figure 7 shows the infrared spectrum of whole live bacteria in the 1000-2300 cm^{-1} region. While the 1000-1800 cm^{-1} region contains a variety of bands from lipids, proteins and other biomolecules, the C-D stretching bands of the endogenous fatty acid acyl chains occur between 2000-2200 cm^{-1}, a region free from interferring bands. Since the deuterated fatty acids are incorporated as lipid acyl chains, these bands provide a specific probe of the thermal response of membranes in live bacteria. Similar to the CH_2 stretching modes, the frequencies of the CD_2 stretching bands of acyl chains depend on the degree of conformational disorder and hence can be used to monitor the average trans-gauche ratio in the lipid ensemble (6-8, 33-36). A frequency of 2090 cm^{-1} for the symmetric CD_2 stretching band is characterisitic of a highly ordered gel phase with a low gauche population. Higher frequencies, the upper limit being about 2096 cm^{-1}, are indicative of an increased population of gauche rotamers. However, the relation between frequency and conformational disorder is not linear. This necessitates the use of a simple two-component overlapping band model to obtain the proportions of gel and liquid crystal phases (26).

Figure 8 shows the temperature profiles of the CD_2 symmetric stretching band in the spectra of the live cells and isolated membranes. Below 20°C, membranes of live cells are in the gel phase; between 20 and 34°C they undergo a transition to the liquid crystalline phase. On cooling the system reverts to the gel phase with a slight hysteresis.

The isolated membranes also undergo a gel to liquid crystal phase transition. However, while the widths of the transitions are the same for isolated membranes and live bacteria, the transition of the former occurs at a temperature about 4°C higher. Thus, at the growth temperature of the bacteria, i.e. 30°C, the liquid crystalline phase content is only about 20 percent, as compared to 50 percent in the live cell membranes. The "fluidity" of biological membranes at a given temperature, which is dependent on the amount of conformational disorder in the lipid bilayer, i.e., the liquid crystalline phase content, is an important factor in membrane regulation.

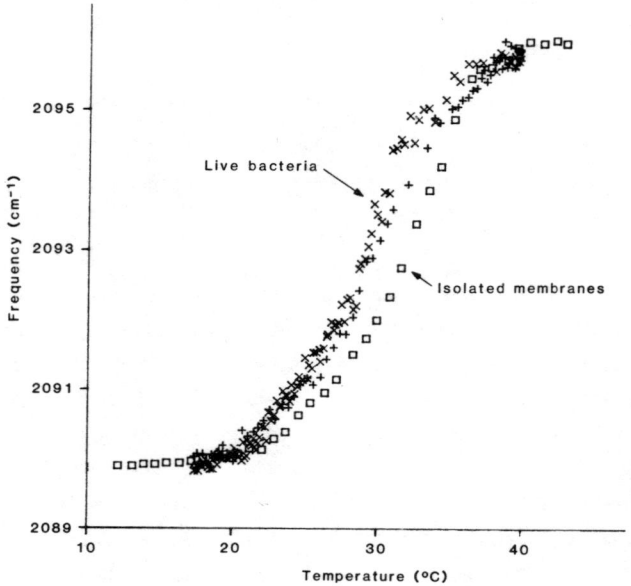

Fig. 8. Temperature dependence of the frequency of the symmetric CD_2 stretching band of the lipids of Acholeplasma laidlawii B grown at 30°C on perdeuteromyristic acid. Shown are frequencies from spectra of live cells with the temperature ascending from 20 to 40°C (+) and descending from 40 to 15°C (x) and frequencies from spectra of isolated membranes with the temperature ascending from 5 to 45°C ( ).

REFERENCES

1. P.R. Griffiths, Chemical Infrared Fourier Transform Spectroscopy, Wiley and Sons, New York, 1975, 340 pp.
2. A.G. Marshall (Ed.), Fourier, Hadamard and Hilbert Transforms, Plenum Press, New York, 1982, 562 pp.
3. J.R. Ferraro and L.J. Basile (Eds.), Fourier-Transform Infrared Spectroscopy, Vols. 1-3, Academic Press, New York, 1982.
4. D.G. Cameron, H.L. Casal and H.H. Mantsch, J. Biochem. Biophys. Methods, 1 (1979) 21-36.
5. G. Weissmann and R. Claiborne (Eds.), Cell Membranes, HP Publishing Co., New York, 1975, 283 pp.
6. H.L. Casal, I.C.P. Smith, D.G. Cameron and H.H. Mantsch, Biochim. Biophys. Acta, 550 (1979) 145-149.

7. H.L. Casal, D.G. Cameron, I.C.P. Smith and H.H. Mantsch, Biochemistry, 19, (1980) 444-451.
8. H.L. Casal, D.G. Cameron, H.C. Jarrell, I.C.P. Smith and H.H. Mantsch, Chem. Phys. Lipids, 30 (1982) 17-26.
9. D.G. Cameron, A. Martin and H.H. Mantsch, Science, 219, (1983) 180-182.
10. D.G. Cameron and R.N. Jones, Applied Spectrosc., 35 (1981) 448.
11. D.G. Cameron and G.M. Charette, Applied Spectrosc., 35, (1981) 224-225.
12. D.G. Cameron, J.K. Kauppinen, D.J. Moffatt and H.H. Mantsch, Applied Spectrosc., 36 (1982) 245-250.
13. J.K. Kauppinen, D.J. Moffatt, H.H. Mantsch and D.G. Cameron, Applied Spectrosc., 35 (1981) 271-276.
14. J.K. Kauppinen, D.J. Moffatt, D.G. Cameron and H.H. Mantsch, Applied Optics, 20 (1981) 1866-1879.
15. J.K. Kauppinen, D.J. Moffatt, H.H. Mantsch and D.G. Cameron, Analytical Chemistry, 53 (1981) 1454-1457.
16. D.G. Cameron, H.L. Casal, E.F. Gudgin and H.H. Mantsch, Biochim. Biophys. Acta, 596, (1980) 463-467.
17. D.G. Cameron, H.L. Casal and H.H. Mantsch Biochemistry, 19 (1980) 3665-3672.
18. J. Umemura, D.G. Cameron and H.H. Mantsch, Biochim. Biophys. Acta, 602 (1980) 33-44.
19. D.G. Cameron, E.F. Gudgin and H.H. Mantsch, Biochemistry, 20 (1981) 4496-4500.
20. H.H. Mantsch, A. Martin and D.G. Cameron, Biochemistry, 20 (1981) 3138-3145.
21. H.H. Mantsch, A. Martin and D.G. Cameron, Can. J. Spectroscopy, 25 (1981) 79-84.
22. R. Mendelsohn, R. Dluhy, J. Taraschi, D.G. Cameron and H.H. Mantsch, Biochemistry, 20 (1981) 6699-6706.
23. D.G. Cameron and H.H. Mantsch, Biophysical J., 38 (1982) 175-184.
24. C. Huang, J.R. Lapides and I.W. Levin, J. Am. Chem. Soc., 104 (1982) 5926-5930.
25. H.H. Mantsch, S.C. Hsi, K.W. Butler and D.G. Cameron, Biochim. Biophys. Acta, 728 (1983) 325-330.
26. R.A. Dluhy, R. Mendelsohn, H.L. Casal and H.H. Mantsch, Biochemistry, 22 (1983) 1170-1177.
27. H.H. Mantsch, A. Garg and D.G. Cameron, Spectroscopy Int. J., 2 (1983) 88-96.
28. H. Sapper, D.G. Cameron and H.H. Mantsch, Can. J. Chem., 59 (1981) 2543-2549.
29. J. Umemura, D.G. Cameron and H.H. Mantsch, J. Colloid Interface Sci., 83 (1981) 558-568.
30. D.G. Cameron, J. Umemura, P.T.T. Wong and H.H. Mantsch, Colloids and Surfaces, 4 (1982) 131-145.
31. H.H. Mantsch, D.G. Cameron, P.A. Tremblay and M. Kates, Biochim. Biophys. Acta, 689 (1982) 63-72.
32. S.C. Chen, J.M. Sturtevant and B.J. Gaffney, Proc. Nat. Acad. Sci. US, 77 (1980) 5060-5063.
33. H.L. Casal, H.H. Mantsch, D.G. Cameron and B.P. Gaber, Chem. Phys. Lipids, 33 (1983) 109-112.
34. S. Sunder, D.G. Cameron, H.H. Mantsch and H.J. Bernstein, Can. J. Chem., 56 (1978) 2121-2126.
35. I.C.P. Smith and H.H. Mantsch, Trends Biochem. Sciences, 4 (1979) 152-154. 154.
36. D.G. Cameron, H.L. Casal, H.H. Mantsch, Y. Boulanger and I.C.P. Smith Biophysical J., 35 (1981) 1-16.

ACCESSORIES AND DATE PROCESSING TECHNIQUES

Different Accessories,
Main Applications and Handling Techniques

K. Krishnan
Digilab Division of Bio-Rad Laboratories

Introduction

With the advent of the commercial FT-IR instruments, and computer techniques, it is now possible to record the infrared spectrum of almost any material regardless of its shape or form. A number of different sampling accessories are available for recording the infrared spectra. Some of these accessories such as ATR and specular reflectance have been used successfully with dispersive instruments, but the FT-IR instruments allow these accessories to be used more rapidly and with greater sensitivity. Most of the sample handling techniques have been reviewed in detail in the series of volumes on "Fourier Transform Infrared Spectroscopy" edited by J.R. Ferraro and J.R. Basile (1). In this paper, some of these techniques will be reviewed with particular emphasis on reflectance techniques (ATR and diffuse) and photoacoustic spectroscopy. Further applications such as far-IR, diamond cell, the absorption subtraction methodology can be found in the article by Krishnan and Ferraro (2). Unless otherwise mentioned, all of the spectra shown in this paper were recorded using the Digilab FTS-IR instruments.

Attenuated Total Reflectance (ATR)

The ATR or attenuated total reflectance (also known as MIR or multiple internal reflectance) technique is a powerful method for the study of the spectra of surfaces. The technique can also be used in cases when only a small thickness of the material under study such as a thick polymer plate or aqueous solution is to be sampled. The samples studied by the ATR technique can be free standing films, coatings on metals or other substrates, powders or liquids. The ATR technique was originally developed by Fahrenfort (3) and has been thoroughly reviewed by Harrick (4). The ATR technique is briefly reviewed here.

If a beam of radiation that is propagating through a prism whose index of refraction is n_1 strikes the surface of the prism at an angle $O = O_c$ (Figure 1), known as the critical angle (such that $SinO_c = O$, where n_2 is the index of refraction outside the prism surface), it is totally reflected.
If the medium outside the prism is totally non-absorbing the reflection is total. If however an absorbing sample is placed in contact with the prism surface, part of the

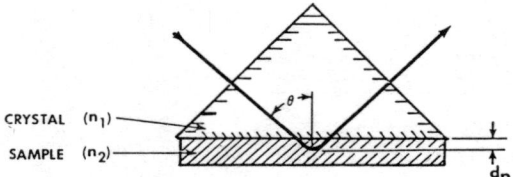

Fig. 1. Schematic of the attenuated total internal reflection process.

incident radiation will be absorbed and the total reflection will be attenuated. At the crystal-sample interface a standing wave is established and there is some penetration of the incident radiation into the sample. The intensity of the radiation that penetrates into the rarer medium falls off exponentially (Figure 2) from its value E_0 at the surface.

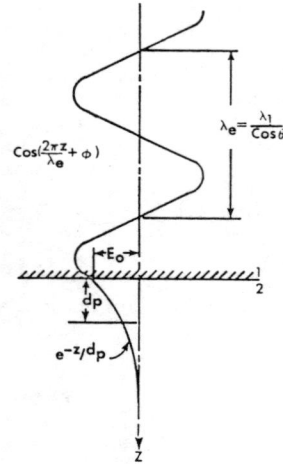

Fig. 2. Definition of the depth of penetration.

It is convenient to define a penetration depth as the depth dp at which E_0 has been reduced to (1/e) of its value at the surface   (The radiation actually penetrates much deeper into the sample.) This depth of penetration is a function of the wavelength λ of the incident radiation, the indices of refraction of the prism and the sample, and the effective angle of incidence. The latter is given as

$$\Theta = \beta + \sin^{-1}\left[\frac{\sin(\alpha-\beta)}{n_1}\right]$$

Here α is the external angle (angle at the baseplate of the ATR accessory) and β is the angle of the crystal endface. The depth of penetration is given as

$$dp = \frac{\lambda}{2\pi n_1 [\sin^2\theta - n_{21}^2]^{1/2}}$$

where n_2 is the differential index of refraction between the prism and the sample.

The sensitivity of the ATR technique can be enhanced by using a long crystal and allowing multiple reflections within it. A typical crystal (50mm long with its endfaces cut at the crystal angle) (Figure 3) one can get between 20 and 30 reflections.

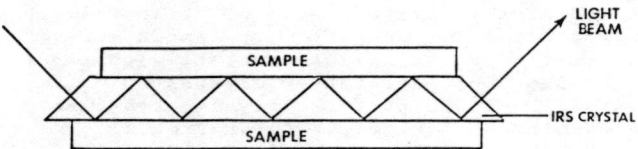

Fig. 3. Multiple internal reflection.

If ℓ is the crystal length and d its thickness, the number of reflections is given as

$$N = \frac{\ell}{d} \cot\theta$$

The ATR accessories designed for use with dispersive instruments with their rectangular slit-image will not function effectively when used with FT-IR instruments, since the latter produce infrared beams with circular cross-sections. ATR accessories specifically designed for use with the FT-IR instruments make use of the circular beam geometry and usually include beam condensing optics. Figure 4 shows the schematic representation of such an accessory designed for use with the Digilab FT-IR instruments.

Table 1

Depths of penetration (dp) in   for Polyethylene

Crystal	Angle	(2000)	(1700)	Wavenumber (cm-1) (1000)	(500)
KRS-5	45	1.00	1.18	2.00	4.00
KRS-5	60	0.55	0.65	1.11	2.22
G_E	30	0.60	0.71	1.20	2.40
G_E	45	0.33	0.39	0.66	1.32
G_E	60	0.25	0.30	0.51	1.02

The materials most commonly employed as the ATR prisms are
KRS-5, germanium, zinc selenide, silicon, silver bromide,
synthetic sapphire and diamond (the latter being used
mainly in the far infrared region). In Table 1 are shown
the typical depths of penetration of the infrared beam for
germanium and KRS-5 of different geometries, when the
sample used is polytehylene.

From this table, it can be seen that by employing a range
of crystals of different geometries and refractive
indices, the depth of penetration can be varied over a
wide range and so the ATR technique can be used for depth
profile studies.

Experimentally, certain precautions are needed to produce
good quality ATR spectra. The surfaces of the ATR prism
should be clean and highly polished-dull surfaces will
result in loss of optical throughput. The sample under
study should make intimate and uniform contact with the
crystal, since the depth of penetration depends on
these factors as well. It is particularly important to
maintain the uniformity of contact if further data
processing, such as absorbance subtraction, and
quantitative measurements are to be made from the ATR
spectra. Reproducible contact between the crystal and the
sample can be ensured by employing a torque wrench set at
a constant value to press the sample against the crystal.
From the above discussion, it is also obvious that the
sample should be pliable, and have a smooth surface.
Since the FT-IR instruments are single beam instruments
(any spectrum actually recorded is a single beam emission
curve of the source modified by the instrument and the

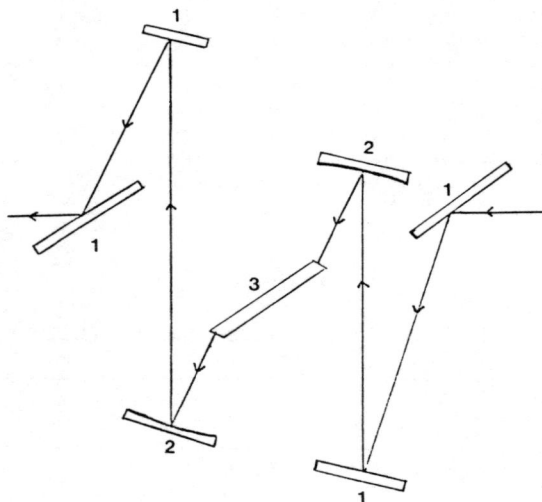

Fig. 4. Schematic of a commercial FT-IR ATR accessory.
(1) Flat mirrors; (2) parabolic mirrors and (3) the ATR crystal.

sample), the actual procedure of recording the ATR spectra involves recording a background spectrum with only the ATR crystal, and a spectrum with the sample being placed against the crystal and ratioing the latter to the former spectrum. Whereas good quality spectra can be obtained with the normal room temperature DTGS detector of the FT-IR instruments, greatly enhanced sensitivity can be obtained by using a liquid nitrogen-cooled MCT detector. Under these conditions, the ATR spectra could routinely be obtained in less than a minute of measurement time. For most samples, KRS-5 crystals can be employed since they have large depths of penetration and will produce infrared spectra over the entire mid-IR (4000-400 cm-1).

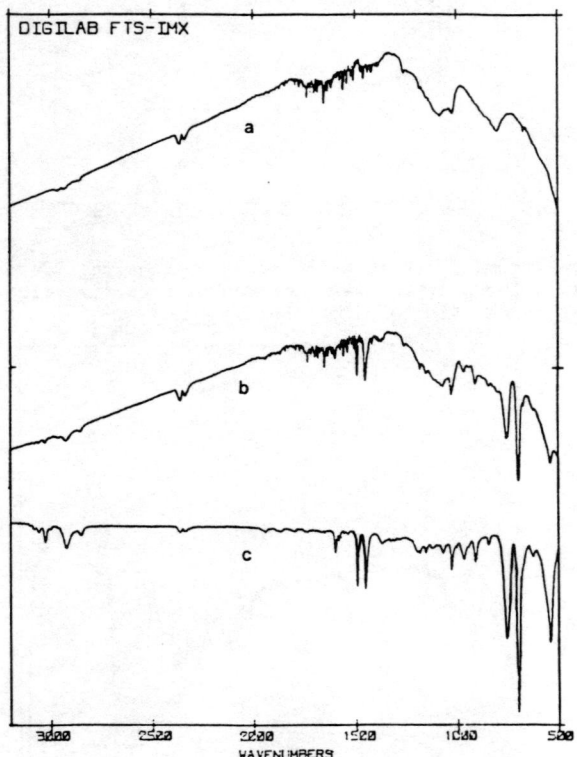

Fig. 5. The single beam and transmission spectra of polystyrene. Using a KRS-5 45° crystal. (a) Reference; (b) sample; (c) ratioed spectrum. The spectra were recorded at 4 cm^{-1} resolution using a MCT detector.

Figure 5 shows the single beam (background and the sample) and the ratioed spectra of polystrene film recorded in 30 seconds at 4cm-1 spectral resolutions. The ATR technique can be used effectively for obtaining the FT-IR spectra of difficult samples such as carbon-filled polymers. Figure

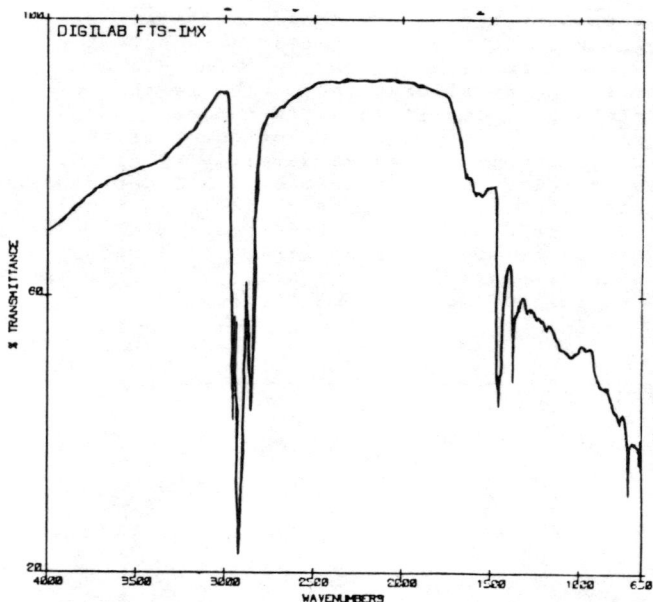

Fig. 6. The ATR spectrum of a carbon-filled polymer using a Germanium 45° crystal. The spectrum was recorded at 2 cm^{-1} resolution for a measurement time of 3 minutes.

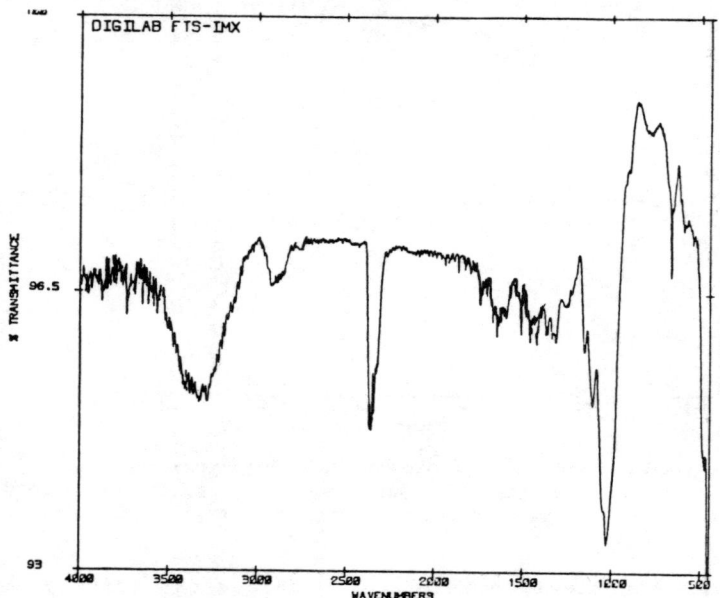

Fig. 7. The ATR spectrum of a 1 mm diameter dark spot from a paper stock. Spectrum was recorded using a KRS-5 45° crystal. Measurement time was 3 minutes.

6 shows the ATR spectrum of a carbon-filled polymer sample recorded using a germanium crystal.

The samples used in recording the above spectra were two 50mmx3mmx2mm pieces. The ATR technique can also be used to record the spectra of very small amounts of sample, when only a single reflection may take place over the sample area.

Figure 7 shows the ATR spectrum of an imperfection in a paper stock whose dimensions were of the order of mmx1mm.

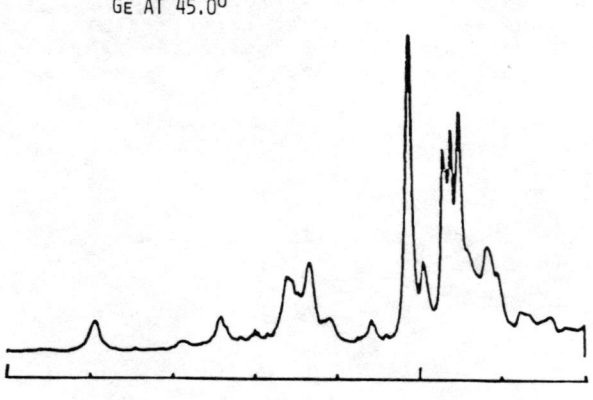

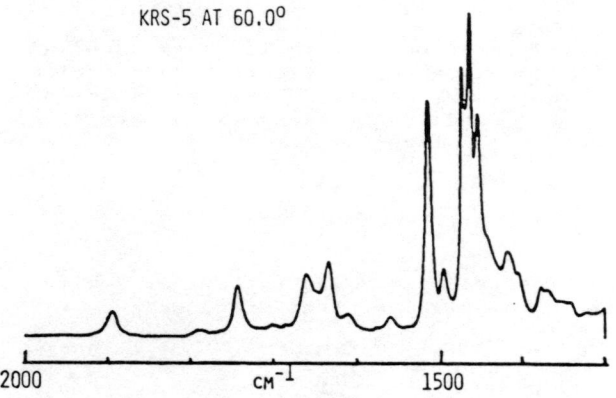

Fig. 8. Comparison of the ATR spectra of a polyethylene sample coated with poly-para-xylylene recorded using a Germanium (45°) and KRS-5 (60°).

Figure 8 shows an example of the depth-profile study using the ATR technique. The sample was vapor deposited poly-para-xylylene on polyethylene. The Figure shows the ATR spectra of the sample recorded using a germanium (45°) crystal and a KRS-5 (60°) crystal. These last spectra were recorded by Smith (5), where more details of the ATR depth-profile experiments could be found.

The use of a polarizer with the ATR technique will allow one to study the molecular orientation of surfaces. Any solid sample will have three optical constants k_x, k_y, and k_z. For isotropic samples, all three optical constants will be identical. For uniaxial samples, two of the constants will be the same and differ from the third one. For biaxial samples, all three optical constants will be distinct. Flournoy and Schaffers (6) have developed the theoretical basis for the ATR from anisotropic media. Sung (7) has described a modified ATR dichroic technique whereby most of the experimental uncertainties could be minimized. Suppose, the sample under study is a polymer with its draw axis the x-axis (Figure 9).

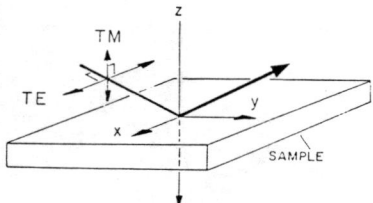

Fig. 9. Definition of parameters for polarized ATR experiments.

Then the sample could be oriented with its x-axis normal to the plane of incidence, and two spectra recorded with the polarizer rotated by 90° degrees. These would define the reflectivities R_{TE_x} and R_{TM_x} (transverse electric and transverse magnetic). Then

$$\ln R_{TE_x} = -\alpha k_x$$
$$\ln R_{TM_x} = -\beta k_y - \gamma k_z$$

When the sample is rotated by 90° about the z-axis with the y-axis normal to the plane of incidence, we get

$$\ln R_{TE_y} = -\alpha k_y$$
$$\ln R_{TM_y} = -\beta k_x - \gamma k_z$$

Here α, β, and γ are functions of the refractive indices of the crystal, the sample and the angle of incidence. Thus, experimentally one can record four spectra and record the three optical constants. Krishnan et al (8) have recorded the dichroic ratio of uniaxial PET using this technique and have compared the dichroic ratios so obtained with those obtained from polarized photoacoustic measurements.

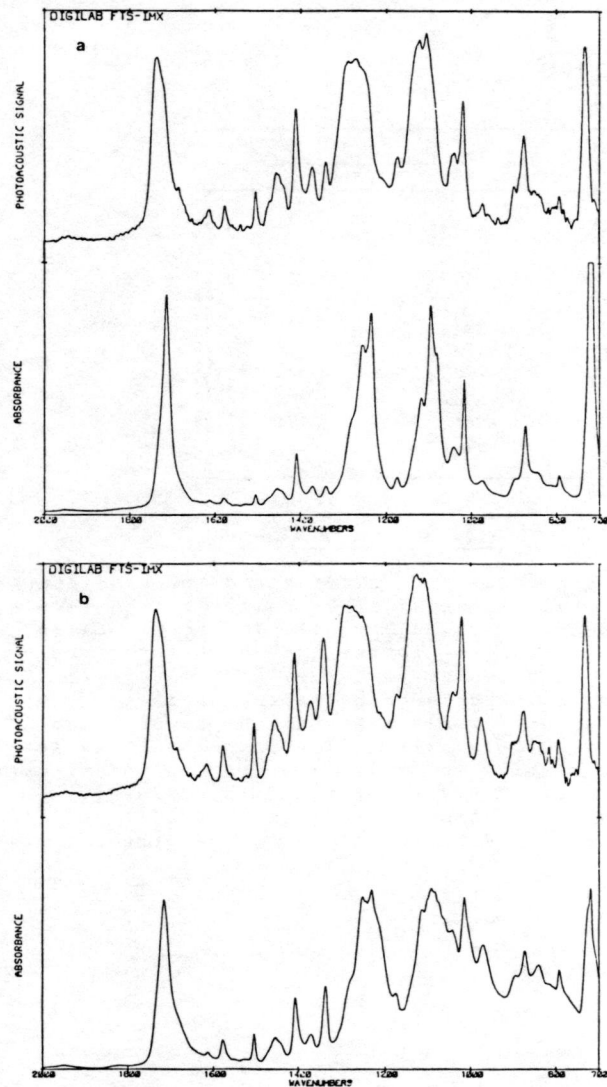

Fig. 10. Polarized ATR and Photoacoustic (PA) spectra of uniaxial polyethylene terepthalate. (a) k_x and (b) k_y. The spectra are shown at the top in each case.
Note the saturation effects in the PA spectra.

Figure 10 shows the k_x and k_y spectra of uniaxial PET obtained using the two techniques and Table 2 shows the comparison of the corresponding dichroic ratio (k_y/k_y) of PET.

Table 2

Comparison of the Dichroic Ratios of PET Measured Using ATR and PAS.

Cm^{-1}	Dichroic Ratio	
	ATR	PAS
1715	1.30	1.00
1580	0.42	0.67
1505	0.36	0.64
1452	0.70	0.97
1408	0.86	1.08
1370	0.83	1.43
1335	0.17	0.39
1175	0.53	0.93
1040	0.62	0.96
1018	1.09	0.83
975	0.26	0.36
895	0.77	1.03
875	1.39	1.33
845	0.63	0.89
795	0.59	0.72

Diffuse Reflectance

The popular techniques for recording the infrared spectra of solids have been to press a KBr pellet of the sample, or grind the sample in a Nujol mull. The former technique is time consuming and the latter method suffers from spectral interferences due to the mulling material. The technique of diffuse reflectance overcomes these problems. Here, one simply mixes up the sample under consideration with an alkali halide powder such as KBr or KCl. The IR radiation is directed on to the powder matrix by a suitable optical arrangement and is diffusely scattered in all directions. Part of this diffuse radiation is collected and directed towards the detector (Figure 11)

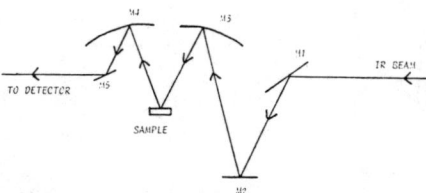

Fig. 11. Schematic optical arrangement for diffuse reflectance.

Diffuse reflectance is a very energy deficient technique, but with the optical through-put capabilities of the FT-IR instrument, spectra can be routinely obtained in very short measurement times.

The technique of diffuse reflectance and its quantitative aspects have been reviewed by Fuller and Griffiths (9). Experimentally, one records a reference spectrum using a transparent and reflective material such as KBr or KCl powder. Then, the sample under study is intimately mixed with the matrix material, and its spectrum recorded. The ratio of the latter to the former spectrum yields the diffuse reflectance spectrum, R. This reflectance spectrum may vary with the particle size of the powders. However, consistent spectra could be obtained if the particle sizes are roughly kept constant throughout a series of experiments. Samples could be used neat (instead of being mixed with the matrix powder), but considerable distortions in relative peak heights within the spectrum may take place if the diffuse reflectance accessory does not completely eliminate the specularly reflected ratiation. If the particle sizes of the neat sample under study are large, the specular component may be considerable.

If the diffuse reflectance technique is to be used for quantitative studies, it will be necessary to convert the reflectance spectrum R into the Kubelka-Munk function $f(R)$
$$f(R)=(1-R)^2/2R$$

The Kubelka-Munk plot of the diffuse reflectance spectrum will be quite similar to the absorbance spectrum obtained in transmission from KBr disks, as can be seen from Figure 12 for a coal sample.

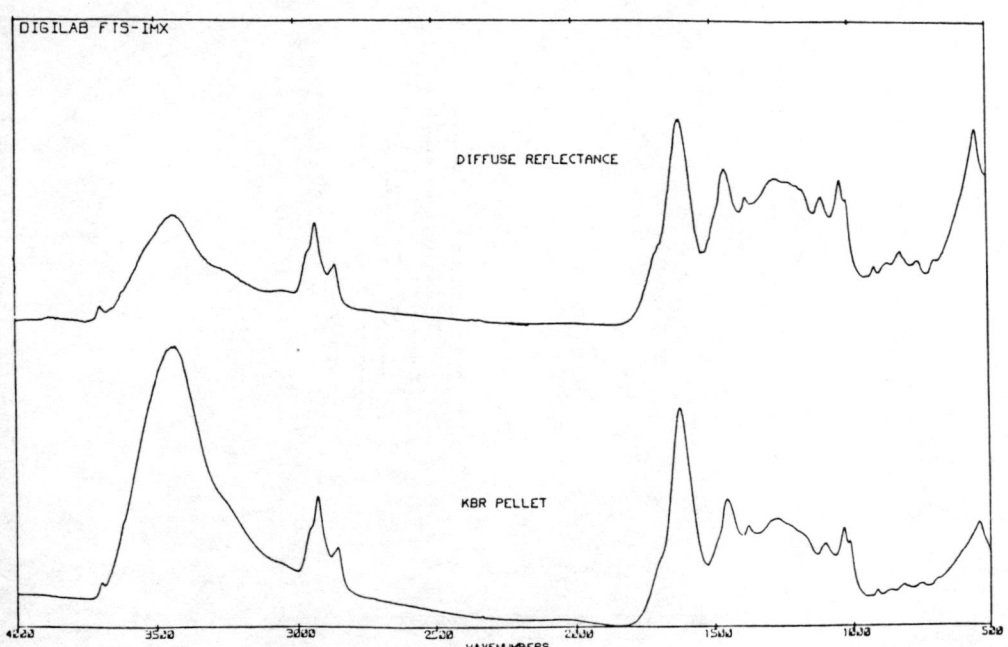

Fig. 12. Comparison between the KBr pellet and diffuse reflectance spectra of a coal sample.

If a liquid nitrogen-cooled MCT detector is used in the FT-IR instrument, the diffuse reflectance spectra could be recorded in a few seconds.

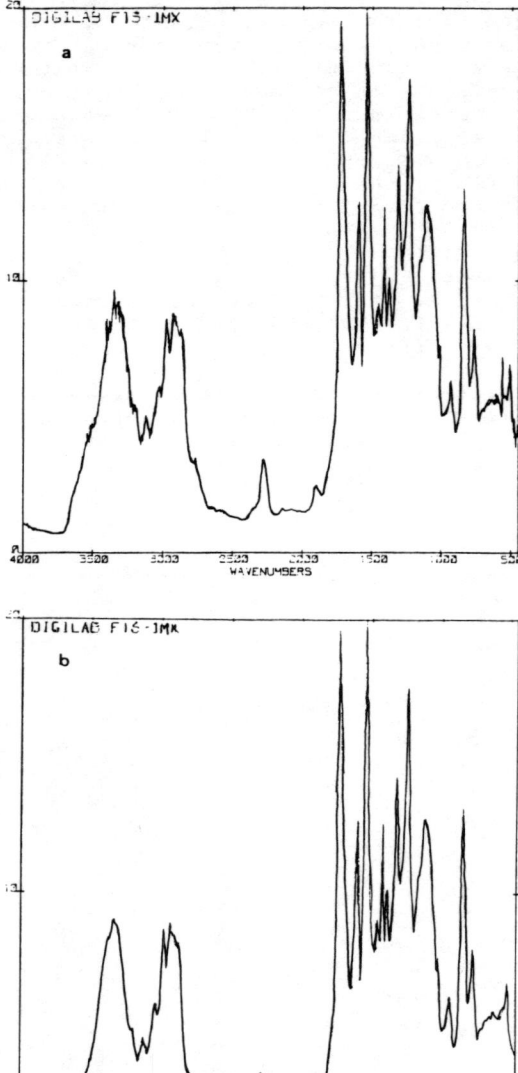

Fig. 13. Diffuse reflectance spectra of a piece of packing foam. An MCT detector was used. (a) Spectrum recorded in two seconds (b) spectrum recorded in 2 minutes.

Figure 13 shows the diffuse reflectance spectra of a piece of packing foam recorded in 2 seconds and in one minute. The diffuse reflectance technique can also be used effectively for the study of solutes dissolved in a volatile solvent, such as HPLC fractions. In this case, a small amount of the solution is deposited on the matrix powder and the solvent allowed to evaporate before the spectrum is recorded. Figure 14 shows the spectrum of two micrograms of Endrin from an HPLC fraction (5% ethylacelate in isooctane being the mobile phase). This technique can also be used for recording the spectra of cold-trapped GC fractions.

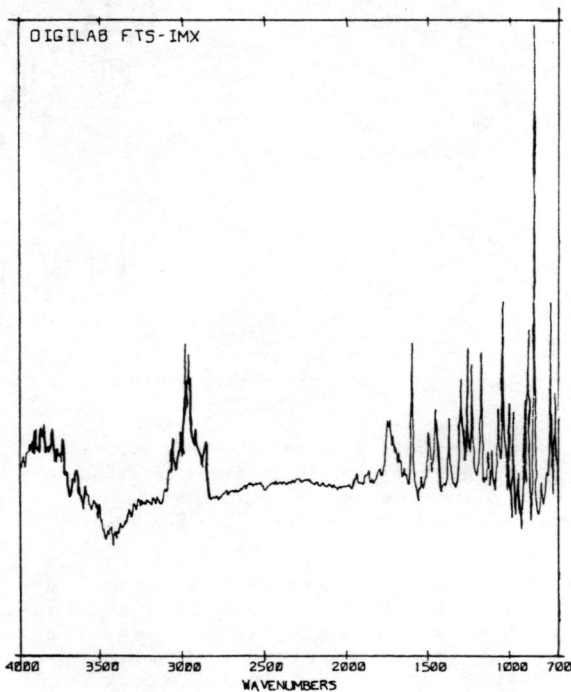

Fig. 14. Diffuse reflectance spectrum of 1 µg of Endrin deposited on KBr powder from an HPLC column. S-1 ethyl acetate in iso-octane was the solvent. Most of the solvent was removed by evaporation at room temperature.

As has been discussed by Fuller and Griffiths (9), diffuse reflectance can also be used as a quantitative technique over a limited concentration range. Absorbance subtraction can be done on diffuse reflectance spectra in the Kubelka-Munk format.

Figure 15 shows an example of such subtraction. Fuller and Griffiths (9) have also shown the feasibility of using the technique for recording the spectra of TLC spots. Another area of application of the diffuse reflectance technique is in following reactions of adsorbed species on

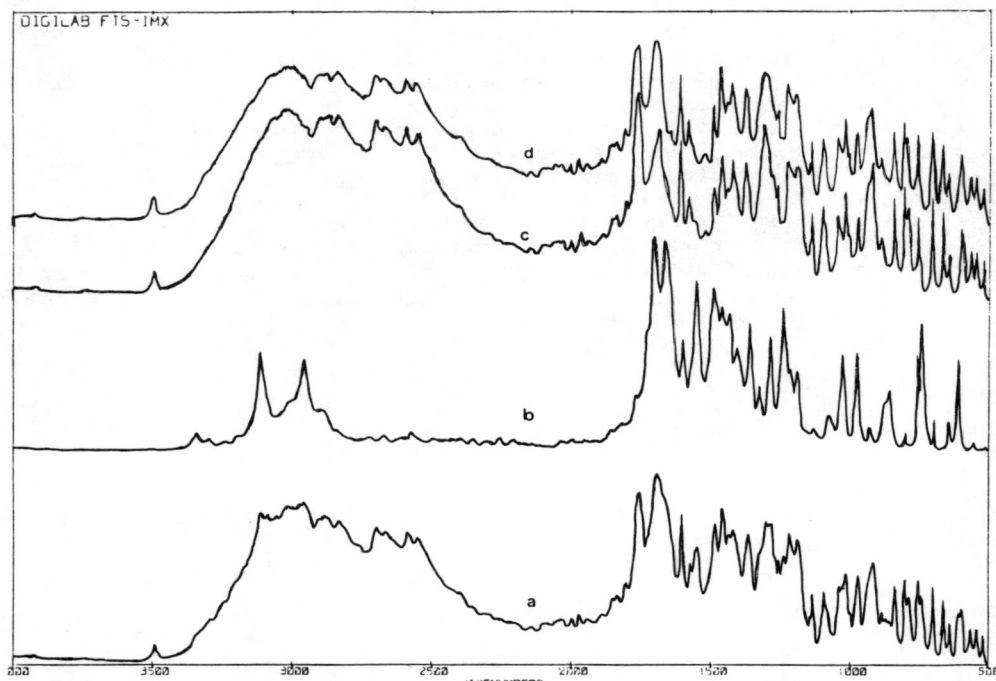

Fig. 15. Absorbance subtraction using Kubelka-Munk functions. (a) Mixture of acetylsalicylic acid and caffeine. (b) Pure caffeine (c) Difference and (d) pure acetylsalicylic for comparison. All samples were kept in a KBr matrix.

catalysts without having to press the catalysts into free-standing thin disks. Using a heatable and evacuable stage for the diffuse reflectance, in situ catalytic studies could be performed (Smyrl 10). Thus, the diffuse reflectance technique can be used in the study of a wide range of samples.

Photoacoustic Spectroscopy

FT-IR photoacoustic spectroscopy (PAS) is rapidly gaining acceptance as a useful technique for the study of gases, liquids and solids. Most of the earlier references to the work on the PAS technique have been reviewed by Krishnan (11). Some of the earlier FT-IR PAS spectra have been presented by Vidrine (12).

The principle of the PAS technique is very simple. The sample under study is placed in a sealed cell of low volume that also contains an inert gas and a sensitive microphone (Figure 16).

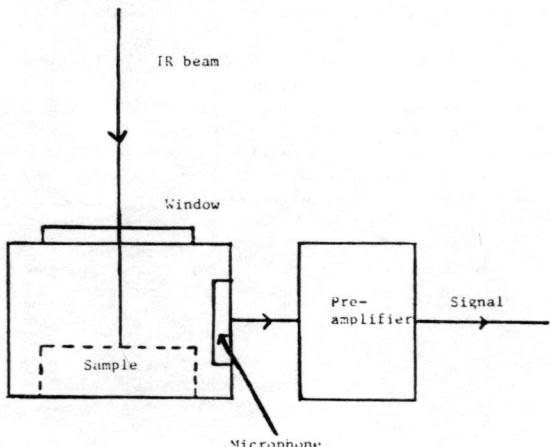

Fig. 16. The principle of photoacoustic spectroscopy.

The IR beam, suitably modulated (by a chopper for dispersive IR instruments or by the Michelson interferometer in the FT-IR instruments) is directed on to the sample. If the sample absorbs the incident radiation at any particular wavelength, the absorbed radiation is converted to heat by non-radiative processes. This heat is transferred to the surrounding gas by thermal diffusion. The gas expands and contracts at the modulation frequency resulting in a pressure wave within the sealed cell. This pressure wave is detected by the microphone and the signal from the microphone then becomes the output of the photoacoustic detector that is processed by the IR instrument detector electronics.

The photoacoustic signal produced is usually a complex function of the incident energy, the absorption coefficient of the sample at a given wavelength, and the thermal properties of the sample. Rosencwaig and Gersho (13) have developed the theory of the PAS. The principles and detailed treatment of the PA technique have been reviewed by Rosencwaig (14) in his book.

It was mentioned earlier that the IR radiation falling on the sample should be suitably modulated. In FT-IR instruments, the Michelson interferometer acts as the modulator. If a monochromatic source of wavelength (wavenumber $\bar{\nu}$) is used as the source in a conventional Michelson interferometer, where the moving mirror is being translated with a constant velocity v, then the interferogram coming out of the instrument will have a modulation frequencey (Griffiths, 15)

$$f = 2 v \bar{\nu}.$$

For the PA experiments, as will be explained shortly, the moving mirror velocity is usually quite small. If we assume this velocity to be 0.16cm/sec (common to all

Digilab instruments), then the modulation frequencies at
1000cm^{-1} and 4000cm^{-1} will be 320 and 1280 H$_z$
respectively. That is, as far as the FT-IR instrument is
concerned, each frequency component of the infrared source
contributes a frequency component to the interferogram.
According to Rosencwaig and Gersho, when dealing with
solid samples, the depth of the sample surface from which
the PA signal emanates is dependent on the modulation
frequency; the lower the modulation frequency, the greater
will be the depth of penetration. Furthermore, in FT-IR
PAS, the depth of penetration will increase with
increasing wavelength (decreasing $\bar{\nu}$). In this respect,
the FT-IR PA spectrum will bear a resemblance to the ATR
spectrum.

The Rosencwaig-Gersho theory yields a complex function for
the photoacoustic signal. This function could be
simplified for six special cases and these are described
below.

Let the sample thickness be l, and the optical path
length $2\beta = 1/\beta$, where β is the optical absorption
coefficient. The thermal diffusion length in the solids
is defined as $\mu = (2K/\rho C \omega)^{1/2}$. K, ρ, C, and ω are the thermal
conductivity, density, specific heat, and the chopping
frequency, respectively. Using these parameters, solids
can be divided into six categories as follows.

1a. Optically transparent and thermally thin solids
($\mu \gg l$, $\mu > 2/\beta$): The photoacoustic signal is proportional
to βl and has a ω^{-1} dependence.

1b. Optically transparent and thermally thin solids
($\mu > l$, $\mu < 2/\beta$): The photoacoustic signal is proportional
to βl and has a ω^{-1} dependence.

For a given infrared absorption band, and a fixed mirror
velocity of the FT-IR instrument (β and ω constant), the
photoacoustic signal will be proportional to the sample
thickness.

1c. Optically transparent and thermally thick samples
($\mu < l$, $\mu \ll 2/\beta$);

The signal is proportional to $\beta \mu$ and varies as $\omega^{-3/2}$. That
is, only the radiation absorbed within the first thermal
diffusion length contributes to the signal, and the signal
is independent of the sample tickness. Since μ has a $\omega^{-1/2}$
dependence, changing the modulation frequency will change
the thickness of the sample contributing to the
photoacoustic signal. In FT-IR spectroscopy one can, in
principle, obtain the spectra of different layers of the
sample by varying the mirror velocity. Examples of such
depth profile studies using FT-IR spectroscopy have been
presented by Vidrine (12)

2a. Optically opaque and thermally thin solids ($\mu \gg l$
, $\mu \gg 2/\beta$). The photoacoustic signal is independent of β
and varies as ω^{-1}.

2b. Optically opaque and thermally thick solids ($\mu < 2$, $\mu < 2\beta$). The signal is independent of β and varies as ω^{-1}.

In case 2a and 2b, since most of the radiation is absorbed with a length that is smaller than the thermal diffusion length, photoacoustic saturation sets in.

2c. Optically opaque and thermally thick samples ($\mu \ll 2$, $\mu < 2\beta$). The signal is proportional to $\beta\mu$ and varies as $\omega^{-\frac{1}{2}}$. As is case 1c, only the radiation absorbed within the first thermal diffusion layer contributes to the signal.

It can be seen from the above discussion that the saturation effects may be important in the appearance of the FT-IR PA spectra. From the theoretical point of view, one can enhance the PA signal by employing lower modulation frequencies (lower mirror velocities in the case of FT-IR instruments). And, one can do a depth-profile study of the sample surface by varying the mirror velocities. It must be remembered, however, that when very low modulation frequencies are employed, it may be difficult to acoustically seal the PA cell against ambient noise; futhermore, the saturation effects may become severe. For depth profiling studies, it must be remembered that this depth in general will vary as $\omega^{-\frac{1}{2}}$ or $f^{-\frac{1}{2}}$. Thus, to change the sampling depth from 10 μm to 1 μm, the mirror velocity of the Michelson interferometer may have to be increased by a factor of 100. For a typical polymer sample such as polymethylmethacrylate, the thermal diffusion lengths are 6.46, 6.88, 8.43 and 11.92 m at spectral frequencies of 1700, 1500, 1000 and 500 cm^{-1}, respectively, when the mirror velocity is 0.16 cm/sec. As was mentioned before, the modulation frequency f, under this condition at $\bar{\nu}$ = 1000 cm^{-1} is 320 Hz. If the diffusion length were to be changed to 1 μm at $\bar{\nu}$ = 1000 cm^{-1}, then the modulation frequency should be $(8.43)^2 \times 320$ Hz = 22.74 KHz. At this high modulation frequency, with the limited amount of energy available from most IR sources, one may not obtain any discernible PA signal.

Notwithstanding the above limitations, the FT-IR PAS technique can be used very effectively for recording the spectra of samples that may be very difficult to obtain otherwise.

Figure 17 shows the spectrum of 3A molecular sieve pellets. The sample could be studied as such with no particular preparation. Of course, the PA signal could be enhanced by grinding the sample into a fine powder to increase the active surface area. Subtraction between PA spectra of comparable samples such as neat organic powders can be performed easily as shown in Figure 18.

One can also utilize the technique to obtain spectra of adsorbed species on catalysts.

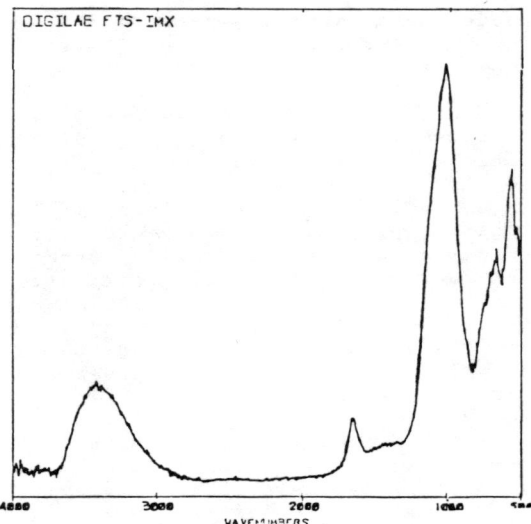

Fig. 17. PA spectrum of 3Å molecular sieves. The spectrum was recorded at 4 cm^{-1} resolution for a measurement time of 8 minutes.

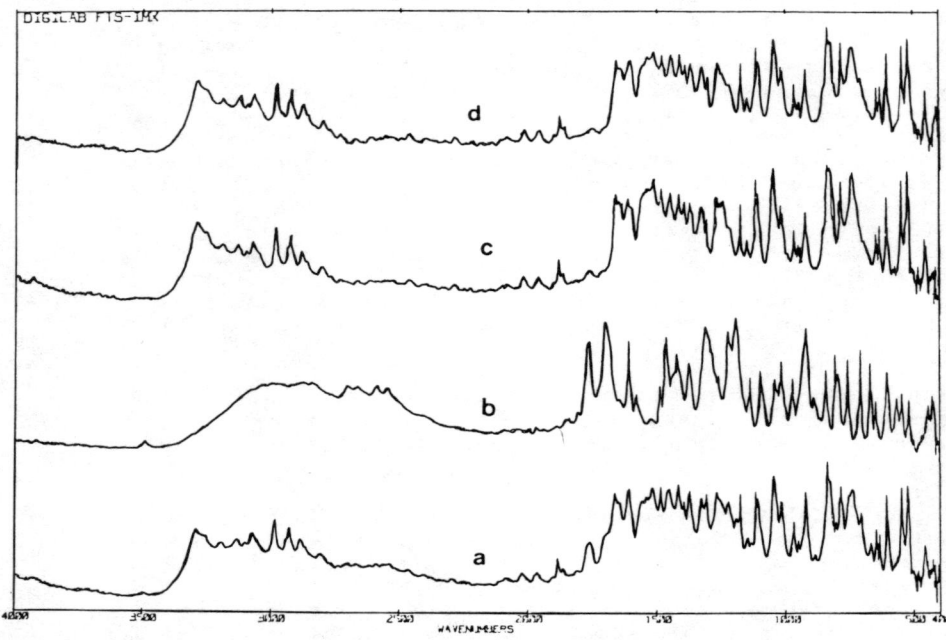

Fig. 18. Subtraction of PA spectra. (a) Mixture of aspirin and phenacetin (b) aspirin (c) difference and (d) pure phenacetin for comparison. The spectra were all recorded at 4 cm^{-1} resolution.

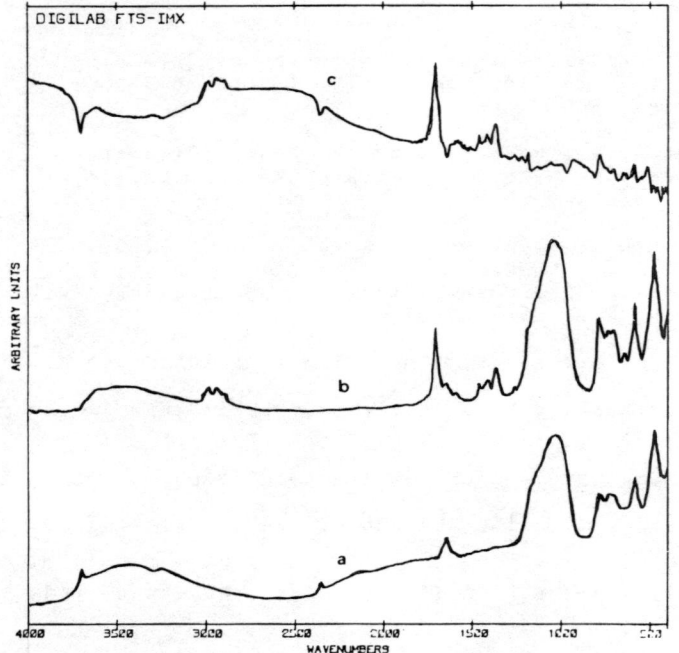

Fig. 19. PA spectra of NaY zeolite. (a) Blank zeolite (b) zeolite exposed to methyl ethyl ketone (c) difference spectrum.

Figure 19 shows the spectrum of a NaY zeolite, with adsorbed methyl ethyl ketone on it. All of the spectra shown here were recorded using a Digilab PA detector, whose schematic representation is shown in Figure 20. Comparison between the FT-IR-PAS spectra and the spectra obtained by other sampling techniques can be found in the article by Krishnan (9)

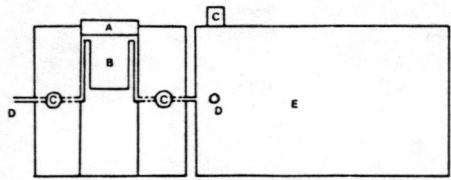

Fig. 20. Schematic of the Digilab photoacoustic accessory.

Acknowledgements

Some of the figures given in this paper have previously been presented in the following references. Figures 1, 2, 3 and 8 - F. Smith (5); Figure 9 - C.S.P. Sung (7); Figure 10 - K. Krishnan et al (8); Figure 11 - K. Krishnan et al., Am. Lab. 12, 104 (1980), Figure 15 - K. Krishnan and J.R. Ferraro ($\overline{2}$); Figures 18, 20 - K. Krishnan (11).

References

1) "Fourier Transform Infrared Spectroscopy" (J.R. Ferraro and L.J. Basile, eds.) Vols. 1-3 (1978-82) Academic Press, New York

2) K. Krishnan and J.R. Ferraro in "Fourier Transform Infrared Spectroscopy" (J.R. Ferraro and L.J. Basile, eds.) Vol. 3 p., 149 (1982)

3) J. Fahrenfort, Spectrochim. Acta. $\underline{17}$ 698 (1961)

4) N.J. Harrick "Internal Reflectaion Spectroscopy" Wiley (Interscience) New York, (1967)

5) F. Smith, Digilab User's Conference (1980)

6) P.A. Fluornoy and W.J. Schaffers, Spectrochim. Acta $\underline{22}$, 5 (1966)

7) C.S.P. Sung, Macromolecules $\underline{14}$, 591 (1980)

8) K. Krishnan, S. Hill, J.P. Hobbs and C.S.P. Sung, Appl. Spectrosc. $\underline{36}$, 257 (1982)

9) M.P. Fuller and P.R. Griffiths, Anal. Chem. $\underline{50}$, 1906 (1978)

10) N.R. Smyrl, Private Communication

11) K. Krishnan, Appl. Spectrosc. $\underline{35}$, 549 (1981)

12) D.W. Vidrine in "Fourier Transform Infrared Spectroscopy" (J.R. Ferraro and L.J. Basile, eds) Vol. 3, 125 (1982)

13) A. Rosencwaig and A. Gersho, Science $\underline{190}$, 556 (1975); J. Appl. Phys. $\underline{47}$, 64 (1976)

14) A. Rosencwaig, "Photoacoustics and Photoacoustic Spectroscopy" Wiley, New York (1980)

15) P.R. Griffiths, "Chemical Fourier Transform Infrared Spectroscopy" Wiley, New York (1975)

DATA PROCESSING TECHNIQUES

D. G. CAMERON

Division of Chemistry
National Research Council of Canada
Ottawa, Ontario K1A OR6

Introduction

This talk will briefly introduce the main data processing techniques available at present, excluding search algorithms. No attempt has been made to cite all the literature, rather, recent references are given and through them the reader can track down other literature.

Data Collection

Resolution should be kept as low as possible. A good guide is that the resolution should be no more than ½ of the full width at half height of the narrowest bands of interest in the spectrum. E.g. If the narrowest band is 16 cm^{-1} wide, use 8 cm^{-1} resolution. The one exception is if you have high SNR and plan to deconvolute, then you can go to 1/3 the width, i.e. in the above case, we would have to run at 4 cm^{-1}. The number of scans must be selected to give the required time, resolution and/or the required SNR.

Precision and Reproducibility

The FT-IR has extremely high reproducibility. That is, if a sample is placed in a spectrometer and several spectra are collected without varying any parameters, the spectra will be reproducible to within hundredths or thousands of a wavenumber. This is, of course, totally independent of the resolution, which only determines how distorted the spectra are. This permits extremely accurate measurements of the positions and widths of bands in the spectra, as the "noise" in the frequency scale is negligible.

In this regard, it should be noted that the "interchange" reproducibility for repeat sampling (i.e. when samples are moved in and out of cells and spectrometers) is lower, and results primarily from optical effects associated with cell construction and placement, and the absolute accuracy is a function of the laser wavelength and spectrometer (including the cell) construction. These figures are generally lower than the repeat scan reproducibility, but the "interchange" reproducibility can approach it if sufficient care is taken.

Frequency and Bandwidth Measurements

As mensioned FT-IR can yield extremely precise frequency and bandwidth measurements. We employ a "centre of gravity" method for frequency measurements, and a simple interpolation routine for bandwidth measurements (1). In brief, the "centre of gravity" technique computes the centre of area of a specified segment of the band. Usually this is the topmost 1-5% of the band. Least squares algorithms supplied by manufacturers work in essentially the same fashion. The most important thing when using these algorithms is to be aware of which part of the band you are using. In Fig. 1a we show a heavily digitized C=O stretching band at several temperatures. When only the topmost 3 data points are used to calculate the frequency a noisy plot is obtained (Fig. 1b). This is because the gradient of the curve over these 3 points is too small. When 11 points are used, the result is much better, because we have now correctly sampled the top of the band.

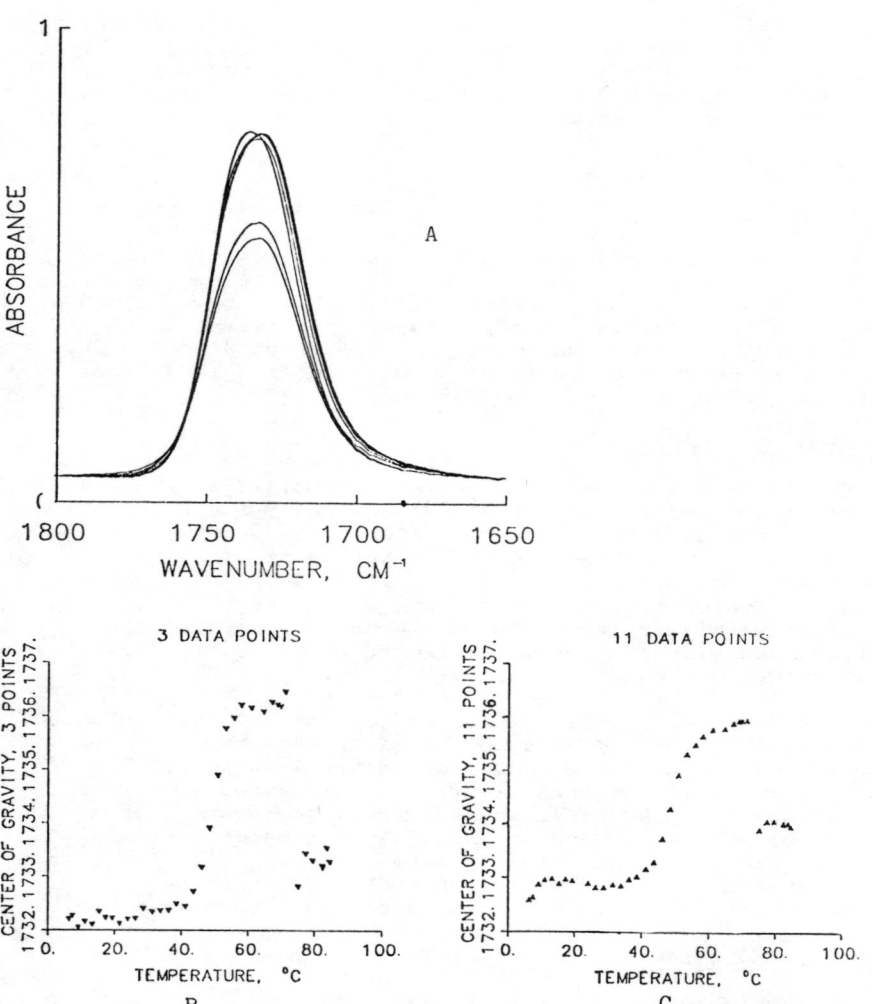

Fig. 1.

DATA PROCESSING TECHNIQUES   161

With regard to bandwidths, consider measuring at points other than the half-height. Fig. 2a shows a band changing with temperature. If we measure the width at half-height, Fig. 2b is obtained. However, if we measure at 8/10ths height, Fig. 2c is obtained. Note that the curve reaches a maximum then decreases slightly. This is strong evidence for the simultaneous presence of two species, i.e. A goes to B, and at some temperatures both A and B are present. If only the half-height were given, this would be missed (see refs. 2 and 3 for further examples).

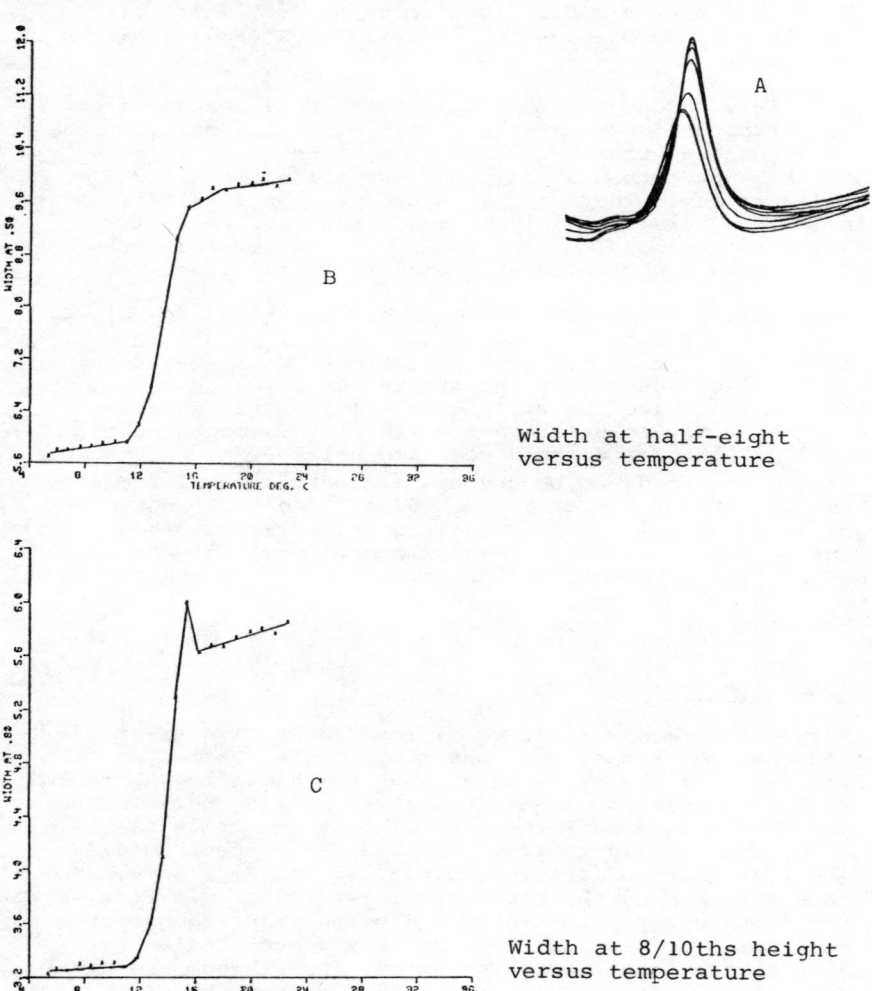

Width at half-eight versus temperature

Width at 8/10ths height versus temperature

Fig. 2.

Deconvolution and derivation

Both of these methods are used to mathematically reduce the width of lines in a spectrum, beyond the level that can be achieved by increasing the resolution.

It should be noted that deconvolution has nothing whatsoever to do with curve fitting, the use of the term deconvolution to mean curve fitting is a misnomer.

The procedures are described in refs. 4-6, which also demonstrate that deconvolution and derivation are variations on a common theme.

In brief, deconvolution removes a known lineshape from your spectrum, and shows what you must mix with that lineshape to obtain your spectrum. The solution is unique. This deconvoluted spectrum will have narrower lines than the original spectrum. They may be asymmetric, which indicates the original lines were truly asymmetric, bands may be split, indicating two or more component bands. The correct integrated intensities and frequencies are retained in the spectrum.

The guidelines for using the method are given in ref. 4. Use of too wide a line for deconvolution will result in negative lobes, use of too little smoothing will result in excessive noise. Generally the process employed is to smooth heavily, and increase the width of the line until negative lobes appear. Having selected the width, the smoothing is decreased until the results become too noisy.

Derivation tends to be used as is, better results can be obtained by appropriate smoothing (5,6). One point not often recognized is that 4th derivatives, or slightly smoothed fourth derivatives can often be successfully employed (5). If your second derivative is noise-free, try going to the 4th.

Some samples of deconvolution are given in Fig. 3.

Difference Spectra

The most commonly employed difference spectra are those obtained when the spectrum of one compound is subtracted from that of a mixture. E.g., a water spectrum might be subtracted from that of a solution. When a sample is being held in the spectrometer while a parameter is being varied it is also useful to sequentially subtract spectra, e.g. spectrum 3-2, 10-5 etc. An example is given in Fig. 4. In Fig. 4a we show a band in a spectrum as the temperature is varied. In Fig. 4b we show the difference spectra obtained by subtracting spectrum n from spectrum n+1, with the spectra having been scaled according to the temperature increment. Apart from the usual demonstration of subtle changes in the spectrum, the form or kinetics of the transition is/are evident.

Band fitting

Band fitting, or curve resolution is a method whereby a least-squares fit of several individual curves to an

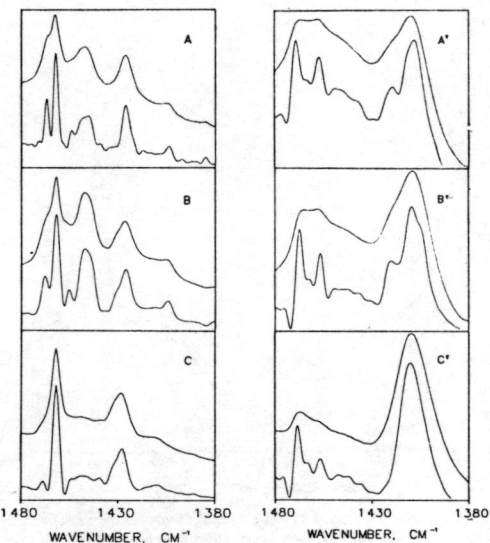

Infrared spectra of the 1480—1380 cm⁻¹ region of coagels of sodium laurate (A), sodium oleate (B), and α,ω-d_3 sodium laurate (C) and of micelles of sodium laurate (A'), sodium oleate (B') and α,ω-d_3 sodium laurate (C'). Each panel shows the experimental spectrum (top) and the resolution enhanced spectrum (bottom) following Fourier self-deconvolution with a Lorentzian line of 6 cm⁻¹ (coagels) or 10 cm⁻¹ (micelles) at full half-width, and smoothed with a Bessel function to K=1.5 (coagels) or K=2.4 (micelles).

DAVID G. CAMERON, JUNZO UMEMURA , PATRICK T.T. WONG, and HENRY H. MANTSCH

Colloids and Surfaces, 4 (1982) 131—145

Casal, Mantsch, Cameron, and Snyder J. Chem. Phys. 77(6), 15 Sept. 1982 2825

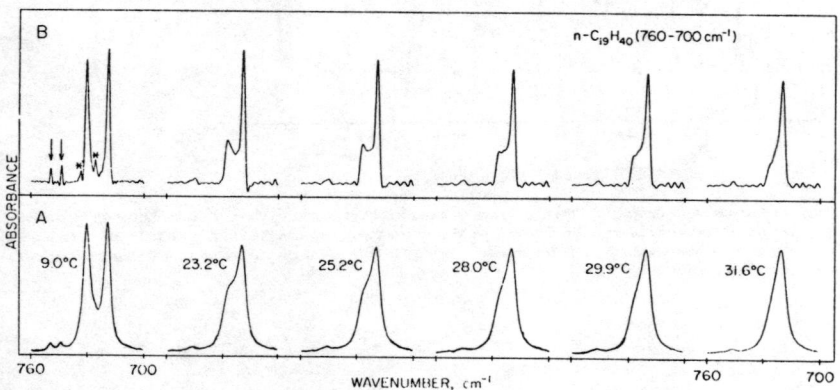

A: 760—700 cm⁻¹ region of the infrared spectrum of n-$C_{19}H_{40}$ at 9°C (phase I) and at 23.2, 25.2, 28.0, 29.9, and 31.6°C (phase II); B: Same spectra as in A after Fourier self-deconvolution, using a 3 cm⁻¹ wide Lorentzian line for the 9°C spectrum, and a 5 cm⁻¹ wide line for the other spectra. The two bands marked with asterisks at 722.5 and 725.9 cm⁻¹ are the components of P_5. The bands at 749.5 and 743.8 cm⁻¹ are the components of P_5; they are marked with small arrows.

Fig. 3.

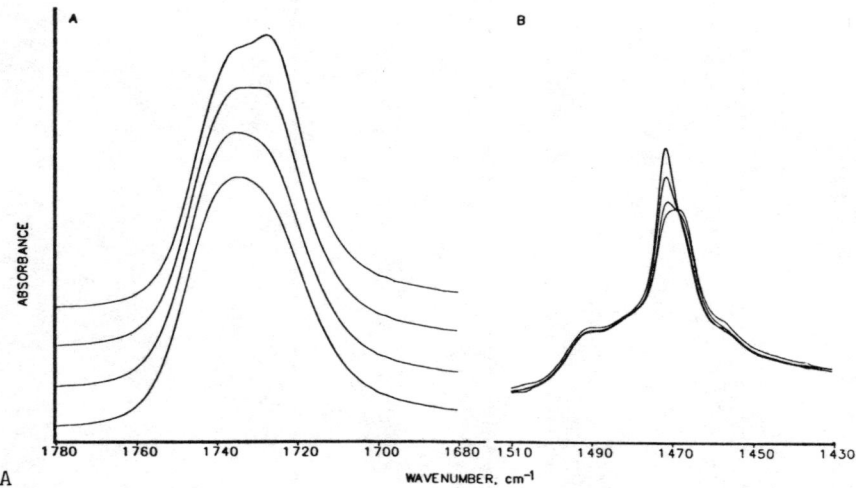

Infrared spectra of DPPC after periods of 2, 24, 48, and 88 h at 2°C. In the region of the maxima, spectra are shown from bottom to top in order of increasing incubation period. *A*, C—O stretching band. Spectra have been displaced relative to each other to permit a better comparison. *B*, CH$_2$ scissoring band superimposed on weak bands resulting from choline and acyl chain methyl groups. Spectra are plotted exactly as recorded, i.e., there is no displacement.

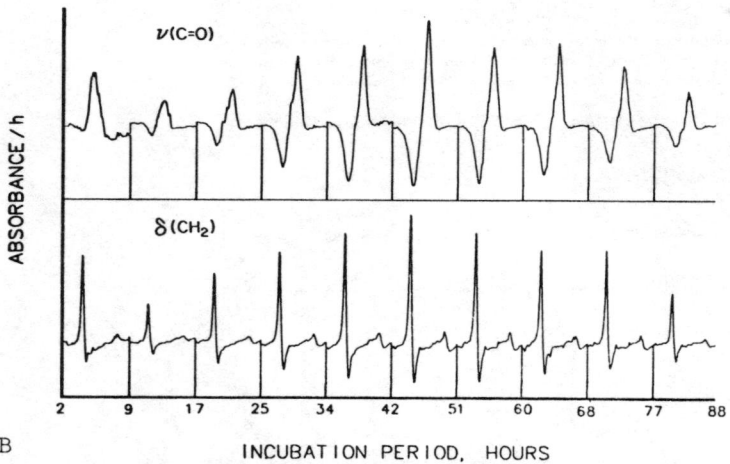

Infrared difference spectra in the regions of the C—O stretching (*top*) and CH$_2$ scissoring (*bottom*) bands during the incubation at 2°C. Spectra have been normalized with respect to the time interval used to generate the spectra. The times used in the subtraction are indicated on the bottom axis; e.g., the left-most difference spectra were generated by subtracting the spectrum recorded after 2 h from that recorded after 9 h, and dividing the result by 7.

Fig. 4.

experimental spectrum is carried out. Generally the curves are Lorentz, or Lorentz-Gauss sum or product functions. One should be extremely cautious when using this method, have an excellent reason for doing it, and tend to disbelieve the results. It should also be remembered that there is no unique solution to this procedure. Some suggestions for the constructive use of the method are to first deconvolute so as to get a realistic indication of the number of component bands, and their positions. You may, however, find that bands are instrinsically asymmetric, in which case you'll have problems fitting with symmetric bands. See ref. 7 for more details.

Factor Analysis

If one can collect a series of spectra in which something (temp., concentration, etc.) is varied, factor analysis can be employed to find out the number of independently varying components in the sample. Details can be found in ref. 8, software is usually supplied by manufacturers.

Multicomponent Analysis

Multicomponent analysis is used to find the concentrations of various species in a mixture, in a sense it is automated spectral stripping although you never see successively substracted spectra. Standard spectra must be available. See ref. 9 for details, manufacturers usually supply software.

Smoothing

Smoothing is a dangerous, but useful, tool. The most popular type of smoothing is least squares, or "Savitzky-Golay" smoothing. Generally smoothing is applicable to broader bands in a spectrum. If the resolution is chosen correctly it can be seen that for the broad bands the bandwidth will be much greater than twice the resolution, (e.g. carbonyl bands are frequently 30 cm^{-1} wide, while CH_2 deformation bands (near 1470 cm^{-1} may be only 8 cm^{-1} wide. A good criterion (10), is to use a cubic smoothing function which encompasses 0.7 times the full-width at half-height (FWHH) of the band to be smoothed, e.g. if a band has 10 data points across its FWHH, use a 7 point smoothing function. A double check is to subtract the smoothed band from the original, if you see anything other than noise, you smoothed too heavily, and are distorting the spectrum.

Be aware that smoothing lowers the resolution. The final resolution is a function of the initial resolution, amount of zero-filling and interpolation, the degree and number of points in the polynomial. It would be simple for manufacturers to calculate and display the resolution of the smoothed spectrum. If they don't, tell them to.

An alternate form of smoothing is Lorentzian smoothing (11). This produces a higher final SNR, but much higher distortion of the spectrum. Be careful.

References

This list is by no means complete, generally one very recent paper is cited, refs. in them should be followed up.

1. D.G. Cameron, J.K. Kauppinen, D.J. Moffatt, and H.H Mantsch, Applied Spectrosc., 36, 245 (1982).

2. J. Umemura, H.H. Mantsch, and D.G. Cameron, J. Colloid Interface Sci., 83, 558 (1981).

3. H. Sapper, D.G. Cameron, and H.H. Mantsch, Can. J. Chem., 59, 2543 (1981).

4. J.K. Kauppinen, D.J. Moffatt, H.H. Mantsch, and D.G. Cameron, Applied Spectrosc., 35, 271 (1981).

5. J.K. Kauppinen, D.J. Moffatt, H.H. Mantsch, and D.G. Cameron, Anal. Chem., 53, 1454 (1981)

6. G.W.F. Maddams, and M.J. Southon, Spectrochim. Acta 38A, 393 (1982).

7. P.C. Painter, R.W. Snyder, M. Starsinic, M.M. Coleman, D.W. Kuehn, and A. Davis. Applied Spectrosc., 35, 475 (1981).

8. J.L. Koenig, and M.J.M. Tovar Rodriques, Applied Spectrosc., 35, 543 (1981).

9. H.J. Kisner, C.W. Brown, and G.J. Kavarnos, Anal. Chem., 54, 1479 (1972).

10. T.H. Edwards, and P.D. Willson, Applied Spectrosc., 28, 541 (1974).

11. R.E. Ernst, in Advances in Magnetic Resonance, Vol. 2, p. 1 (1966), Ed. T.S. Waugh, Academic Press, N.Y. 18.

COMPUTER BASED INFRARED SEARCH SYSTEMS

John P. Coates and Robert W. Hannah

Perkin-Elmer Corporation, Spectroscopy Division
901 Ethan Allen Highway, Ridgefield, CT 06877, USA.

1. INTRODUCTION

The infrared spectra of most chemical compounds are unique and it is this feature that enables a spectroscopist to identify an unknown substance from its spectrum. Traditionally this identification is attempted either by direct interpretation or from a possible match to a standard reference spectrum. A positive match is obviously more desirable but this is usually limited by the availability of a wide range of reference spectra and, if performed manually, it can be a very time consuming operation. The ability of a computer to handle a large number of data files and to be able to perform high speed repetitive operations make it an ideal candidate for spectral searches. In fact, this was considered to be one of the first main applications of the computer for analytical chemistry.

Early work involved data bases compiled from an original ASTM infrared spectral collection that contained spectra from over 100,000 compounds (Ref.1-4). The rules used for encoding the spectral data in these studies has been documented as an ASTM standard (Ref.5) and this was based on the earlier work of Kuentzel (Ref.6). The various programs utilized a range of storage media from punched cards (Ref.1) to magnetic disk (Ref.3/4). Large spectral data bases of this type are now available from time-shared network systems, such as IRGO (Ref.7).

The trend in recent years has been towards dedicated, small computer systems. One of the first applications of a search to a small system based on a desk-top calculator was described by Rann (Ref.8). During the past few years there has also been a resurgence of interest in methods for spectral searching and automated compound identification. This is evidenced by the large amount of published literature on data compression algorithms, data extraction from interferograms and computer based interpretation. A general review of spectral search and retrieval techniques was recently compiled (Ref.9). Automated interpretation is now a popular issue and recent publications have focussed on artificial intelligence methods that involve pattern recognition and/or rule-based techniques(Ref.10-20). Some of the rule-based approaches have attempted to simulate the procedures used by the spectroscopist.

In this presentation we will discuss the practical aspects of computer based search systems. Examples will be included from a commercial search program that incorporates a computer based interpretation.

2. SPECTRAL SEARCH METHODS

As indicated earlier, a relatively large amount of work has been published that describes searching schemes and methods of data compression. Data compression techniques are especially important because in addition to the obvious requirement for accuracy there are also requirements for speed and economy of data storage. These factors are particularly important for small computer systems and usually there is a trade-off between searching efficiency and the degree of data compression. Some of the popular schemes for spectral searching are indicated below :

  a) zone encoding with numeric code comparison,
  b) peak matching with partial, absolute or positive match criteria,
  c) "full" spectral searches with a fully digitized spectrum library,
  d) matching of interferograms or transformed spectra.

Most of these methods have been incorporated into commercial search programs. Individually they have advantages and disadvantages and in many cases the success of a particular method depends on the type of computer, the nature of the samples, the quality of the reference library, the expertise of the operator and finally, how the results are to be utilized. A brief description of the above four methods will now be given.

In the zone encoding method the spectrum is divided into regions or zones and the relative position of the strongest peak within a zone is defined by a digit 0 - 9. These numbers are sequentially combined to give a numeric code for the complete spectrum. This method is very efficient for storage and it allows for high speed searching because only simple numeric comparisons are required. Additional criteria are sometimes added to the encoding, such as the addition of the position of the strongest band. The Sadtler "Spec-Finder" code (Ref.21) is an example of zone encoding.

Peak matching is a popular method because the procedure uses the important components of a spectrum yet it requires relatively little storage overhead. There is an implied assumption that the peaks provide the information necessary to identify a compound. In the past some programs have used partial matching with a limited number of peaks where the most intense are selective. These work well if used by a single operator but they are too subjective for a general procedure. They can also be dependent on sample preparation and instrumental operating parameters. Search methods based on the use of all peaks that satisfy a defined threshold from a preconditioned spectrum usually work well if the program operates to pre-defined criteria. In such cases, if the program is moderately forgiving to variations in sampling and instrument conditions, then it has general application.

"Full" spectrum search methods use, as the name implies, a fully digitized version of the spectrum. When used in this form, the spectra are usually stored at reduced resolution because at full resolution the storage overhead and the time required for processing become excessive. The matching of the unknown spectrum is usually based on a correlation criterion, such as the Dot Product between the library spectrum and the unknown. In practice, the only real advantage of this method is the availability of a digitized spectrum for display purposes.

In recent years workers such as De Haseth and Isenhour have demonstrated that the phase corrected interferogram or the transformed spectrum can be used to advantage for spectral searching. The main feature of this approach is that a relatively small number of points (100 - 200) can be used to adequately represent the spectrum. Again, the degree of match is based on correlation methods. Results appear to be reasonably well discriminating even at high noise levels. It is not possible, however, to accurately reconstruct an original spectrum from the stored segment of data points.

3. APPLICATIONS OF A SEARCH SYSTEM

In an analytical laboratory there are several situations that can be served by computer based search methods.

a) identification of a single unknown,
b) characterization of the components of a mixture,
c) characterization of a complex mixture,
and d) compound/structure/product confirmation.

The first case is usually the one that is considered most. Many algorithms rely on a sample being a single entity to be successful. In practice this is not necessarily always true especially in industrial or forensic laboratories where samples are often complex mixtures. It must be pointed out, however, that a single unknown does not necessarily infer that the sample has to be a single pure compound. It only means that the program treats the sample as a single entity. Therefore if a library is constructed from commercial materials it can still be serviced by a program that identifies a sample as a single unknown. One interesting application of a search program is for the confirmation of the composition or structure of a material. This is especially useful for screening of raw materials and formulated products.

At one time the adage "bigger is better" was used to describe the requirements for spectral libraries. This is really not the case and in fact it is the application that defines the optimum library size. Very large collections of 10K to 150K spectra are generally best for identifying obscure compounds. If they are to be used and maintained efficiently then with the current state-of-the-art they are best used with a time share network. In the future, video disk technology may well change this situation.

In the average laboratory, a large collection containing 2K to 15K well defined spectra is probably very adequate for the general characterization of an unknown. While a library of this size may not always give a positive match, it is likely to yield a reasonable match with a compound that is chemically or structurally similar. A typical example is the EPA vapor phase collection that is used for infrared identification of components separated from a gas chromatograph (GC-IR). If a laboratory works within a specific area of industrial application, such as polymers, lubricants, paints, pharmaceuticals, etc., then smaller specific libraries, 500 to 5000 spectra, are probably more useful. Libraries of this size are necessary if successful mixture component analysis is to be attempted. Mixture algorithms by nature are less discriminating and tend to be less successful when applied to large libraries. Typical examples of specific libraries are the pure polymer collection prepared by Hummel (Ref.22), the drug collections from the Georgia Crime Laboratory (Ref.23/24), the lubricating oil collection from Perkin-Elmer (Ref.25) and the Atlas for the coatings industry (Ref.26). These collections are available as published hard copy (as indicated by the references) as well as digital data.

Only a small number of entries are required in libraries for quality control applications (50 to 200 spectra). It is important, however, that they are carefully prepared by the end-user with the sample and instrument conditions used for the control analysis. The best results for any application will always be obtained from a library prepared by the end-user.

4. SPECTRUM QUALITY CRITERIA

The success of any search program is not only dependent on the nature of the algorithm or the availability of an adequate library but also it is reliant on good, consistent sample preparation. This is particularly the case for library generation by the user. The comments outlined in this section are pertinent for both library development and for the analysis of an unknown. Obviously it is not always possible to follow the ideal procedures and some samples may be more difficult than others, such as micro-samples and samples that require a specialized sampling accessory. If this situation is the rule rather than the exception then a library prepared under these conditions may be more appropriate.

Figures 1/2(a-d) are used to illustrate points that are relevant to ideal sample preparation. Spectrum intensity is one of the most important issues for consistent results. Figure 1 shows a well prepared sample with good band intensities - Figures 2a and 2b are non-ideal extremes. Relative band intensities are the key to a good discrimination and characterization. In the case of the two extremes where a spectrum is either too weak or too strong it is impossible to obtain an accurate representation of weak bands versus strong bands. The case where a spectrum is too strong is especially undesirable because it is impossible to evaluate the importance of an absorption band. Therefore it is a good rule to always ensure that all bands are on scale, i.e. greater than 0% T, with the possible exception of the aliphatic carbon-hydrogen stretching region (3050 - 2800 cm-1). This last point is especially noted for long chain compounds where bands in the fingerprint region are often too weak to be useful if the C-H bands are above 0% T.

If the strongest band in the spectrum is obtained between 10% T and 1% T then it is possible to carry out a software normalization. This sets the spectrum background and the strongest peak to pre-defined values. If this is always carried out then spectra will be compared under almost constant conditions. Also, if the spectra are always normalized then it is possible for the computer to make pre-defined judgements, such as band rejection for weak features and thresholds for automatic peak selection. If a spectrum is not first normalized then any action of this type will lead to inconsistent results depending on sample concentration and thickness.

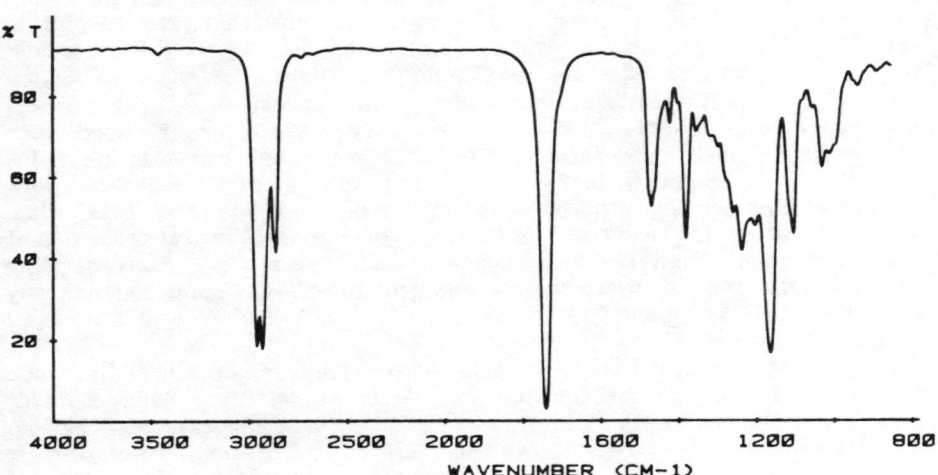

FIGURE 1 : Ideal spectrum for computer based searches.

Normalization is really only meaningful if the spectrum background is free from artifacts. Any background contribution from an accessory or a sampling medium should be removed by absorbance subtraction. Solid samples often have a scatter background superimposed on their spectra. Figure 2c is a simulation of this effect. It is often important to remove this background by software correction.

Some search algorithms are more susceptable to background effects than others. Full spectral searches are affected the most because the indiscriminate use of the full spectrum leads to an inclusion of "mis-information" as well as spectral information from the sample. While removal of the background and spectrum normalization is recommended in all cases it is less important in the case of peak based searches. For example, an experiment with the uncorrected spectrum from Figure 2c gave a 46 % reduction in match quality for a full spectrum search whereas the score for the peak based search was unaffected. The contrived example shown in Figure 2c has a simple linear (in absorbance) slope to the background. The backgrounds superimposed on the spectra from real samples are often more complex and involve high order functions. Simple first order corrections for these backgrounds can produce severe distortions to the overall spectrum and therefore it is important for a search algorithm to be forgiving under these circumstances.

Most high quality spectral collections have been obtained with moderate spectral resolution - 2 cm-1 to 5 cm-1 with a peak-to-peak noise level of less than 1% T. If data are recorded with less resolution then there is a risk of loss of fine detail even for condensed phase spectra. It is this fine detail that helps in the discrimination of chemically or structurally similar compounds. Furthermore, crystalline compounds and also certain liquids, such as indene, can have narrow absorption bands that are resolution sensitive. A high noise level, as indicated in Figure 2d is undesirable but not necessarily unavoidable. Techniques such as signal averaging or digital smoothing (applied modestly) can often help this type of situation. If a peak match algorithm is used then changing the peak pick threshold to a larger value can also be helpful. Some experiments have shown that full spectral searches work quite well under high noise situations. This may be the case when small libraries are used and also when background artifacts are kept to a minimum. High noise conditions often occur with microsampling experiments and in these cases background effects can remove any advantage that is gained.

Good spectrum calibration is also an obvious requirement for good results but this is becoming less critical with modern Fourier transform instruments which generally ensure accurate calibration. Sample preparation techniques can affect calibration, therefore, it is still necessary to be aware of spectrum calibration. Also changes in chemical environment can promote band shifts. If variations are relatively small, such as 0.5 to 5 cm-1, little change is noticed for peak based searches because most can accomodate small shifts.

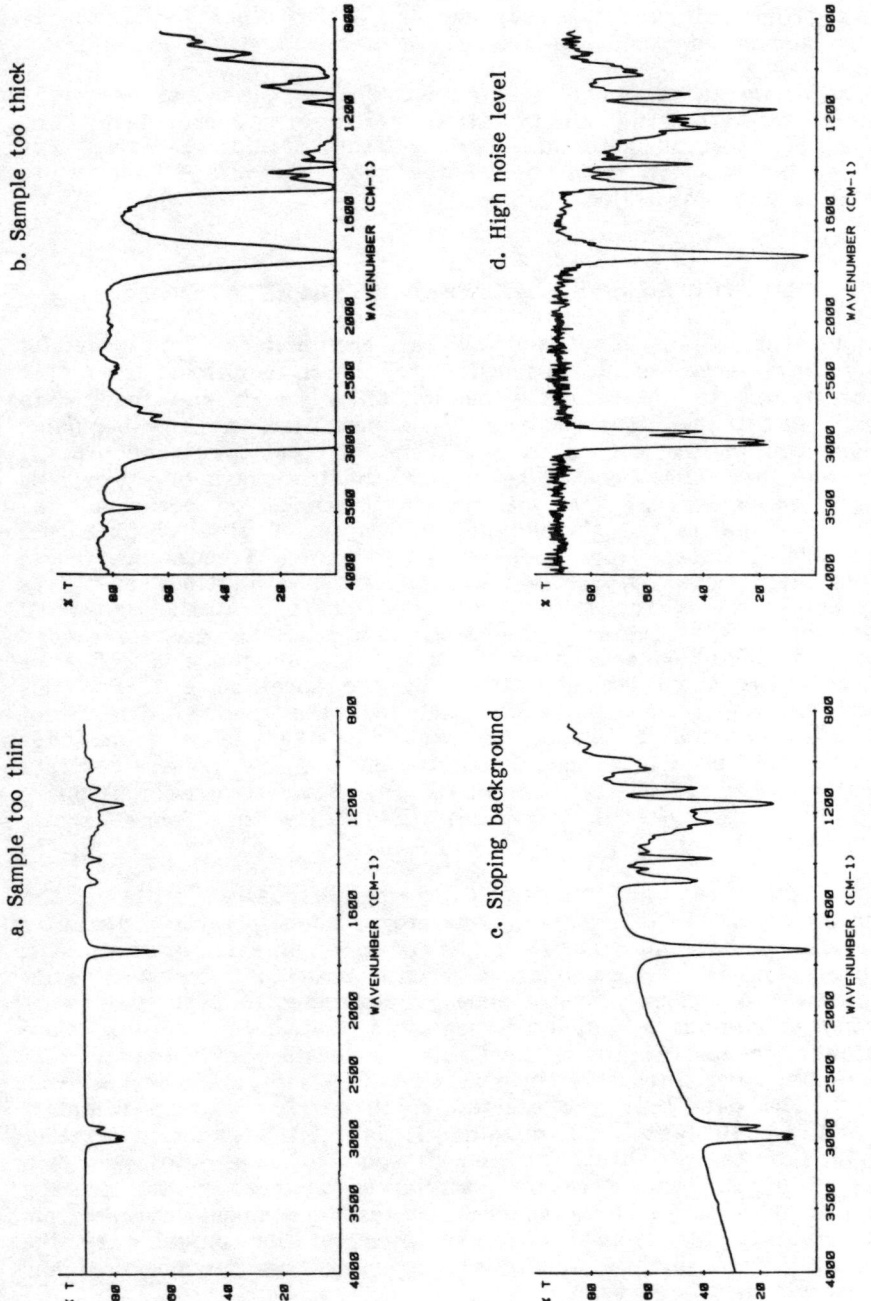

FIGURE 2 : Typical problems encountered in the production of an infrared spectrum.

Results from full spectral searches, on the other hand, can be severely impaired depending on the degree of band shift.

One last comment on sample preparation. The computer has become a powerful tool for the manipulation of spectroscopic data. The removal of background features and artifacts should be carried out with caution and it must be realized that there is no substitute for good sample preparation.

5. COMPUTER ASSISTED SPECTRAL INTERPRETATION

Many of the topics discussed so far are critical for successful results from most search systems. There is an additional step that can be taken to improve the quality of a search and that is to apply a spectrum interpretation as a pre-filter. At the beginning of this article we discussed the role of the spectroscopist. It would be very time consuming if the spectroscopist blindly sorted through reference spectra without any preconceived notion of the basic functionality of the unknown sample. Typically, functional groups are assigned from a knowledge of group frequencies. These are not applied as single values but as combinations of values that pertain to a functional group within a particular chemical or molecular environment. If these combinations are developed as well defined criteria then it is possible to teach a computer to follow the same procedure used by the spectroscopist. For this method to work well it is essential that the spectral data is of good quality and it should be preconditioned by the methods described earlier - i.e. background removal, low noise, moderate resolution and intensity normalization. Without preconditioning the technique may still work but the ability to interpret subtle spectral detail will be lost.

The system featured in the following discussion will be the Perkin-Elmer "SEARCH" program. This program incorporates a computer based interpretation with a series of peak match routines. The interpretation is performed from a data base that contains nearly 900 complex or macro group frequency assignments. Each assignment contains a number of spectral segments that define one or more functional groups within a particular molecular environment. Both wavenumber and intensity information is used to describe each segment. The data base was started in 1977 and took approximately two and a half years to develop. It is relatively general, rather forgiving of sample state and is intended to cover most common classes of organic and inorganic material encountered in the average analytical laboratory. Three volumes of text accompany the program to help qualify the somewhat cryptic interpretation provided by the computer . These include a detailed account of possible chemical and spectroscopic interferences.

The program was written for two different computers - the Model 3600 Data Station, based on the Motorola MC6800, and the Model 7000 series Professional Computer, based on the Motorola MC68000. The optimum use of the system involves direct acquisition of the spectrum into the computer followed by data manipulation to precondition the spectrum and the final automatic generation of a peak table. Facilities are provided to edit unwanted features such as solvent or impurity bands. The user can also define the physical state of the sample and indicate the elemental composition if known.

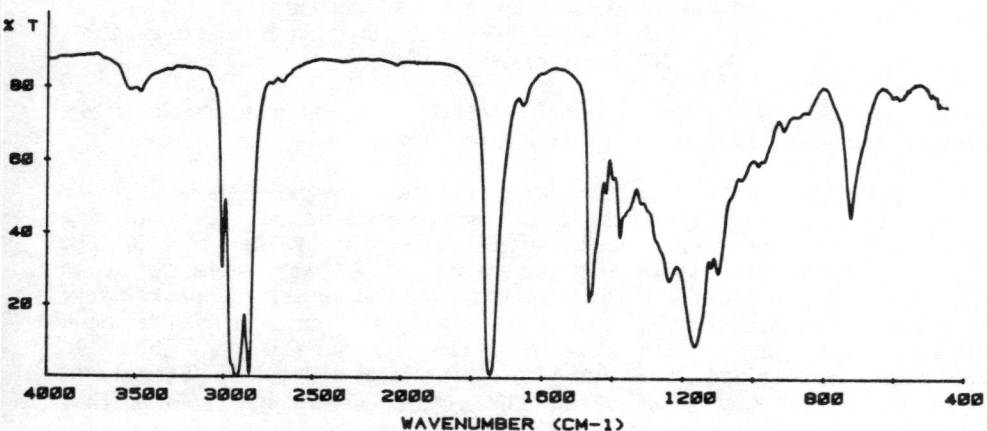

FIGURE 3 : The infrared spectrum of an "unknown" oil, 0.012 mm film between potassium bromide windows.

Once the peak table is entered the interpretion can be made. For some applications this may be all that is required - for example to determine if a material contains a particular class of carbonyl compound. In most cases one of four search algorithms would then be used. These algorithms use the interpretation either completely or partially to bias the search with a weighting for either a single entity or a mixture. Each algorithm is applied concurrently and once completed all four sets of results are available.

The application of this software will be described in the next section for three different analytical scenarios - the identification of a single unknown, the analysis of a binary mixture and the characterization of a complex commercial material.

6. TYPICAL APPLICATIONS OF A SEARCH PROGRAM

The most commonly conceived application of a search program is the identification of an unknown. A typical problem in a service laboratory is the need to analyze a material that has been isolated as a contaminant. The spectrum in Figure 3 could be considered as one such example - an unknown oily material. For a spectroscopist this is an easy example and the diagnosis would be a natural fatty oil, such as a vegetable oil. This would be deduced by the following interpretation steps :

- A) a long aliphatic side chain (2924, 2855, 1465, 1379 and 723 cm-1)
- B) an ester carbonyl (1746, 1237 and 1159 cm-1)
- C) unsaturation , possibly cis configuration (3004 and 1654 cm-1, cis from broadening of 723 cm-1 band)

The net conclusion is an unsaturated long chain aliphatic ester, which is characteristic of a vegetable or animal oil.

The operation with the computer program involves three discrete steps from either the "raw" or the preconditioned spectrum. The computer displays from each step are shown in Tables 1 - 3. The first involves the automatic generation of a peak table based on a 1% T discriminator or threshold. In this case, 29 peaks were defined as indicated by Table 1. Next, the data from the peak table is compared with the the group frequency data base to provide a list of possible functionalities (Possible Structural Units). It is very clear from the results shown in Table 2 that the computer interpretation of the unknown oil is consistent with that obtained from a spectroscopist. In the final stage the peak table is compared to a spectrum library with the interpretation being used as a filter. The results obtained from a comparison with a library of lubricant products (Table 3) support the spectrum interpretation and confirm that the material is a vegetable oil. The first '9' in the scores indicates that all the compounds in the list have similar functionality - i.e. they are all aliphatic esters. The second number in the score is used to define the degree of spectrum match. A result of '9' for the first compound in the list, soybean oil, is indicative of an exact match. The sample used for this illustration was a cooking grade vegetable oil.

A recent application of search programs has been the identification of components in a mixture. The results can be varied and they depend on the nature of the mixture being studied. Mixtures of chemically related compounds can be analyzed relatively easily so long as there are a reasonable number of unique absorption bands for each component. Figure 4 shows the individual spectra of two two barbiturates - barbital and mephobarbital, prepared as potassium bromide pellets. While they both have similar overall spectra they are sufficiently dissimilar to be differentiated. A spectrum from a 1:1 binary mixture of the two drugs is given in Figure 5. This spectrum resembles the original spectra and this is reflected in the computer interpretation - Table 4. The basic structure of a

TABLE 1 : Peak table for "unknown" oil spectrum.

```
X:   3551 4000 -  450     1.00     0.20   100.00 T  F      S3       UNOIL
REF :     4000 98.20  2000    98.99    450
     3531 89.1      3473 88.6    3009 34.3    2924  0.2    2855  0.9
     2730 91.6      2680 92.3    2335 98.8    2030 98.1    1746  1.1
     1657 86.0      1465 24.5    1419 59.3    1398 63.4    1379 45.4
     1239 31.1      1165 10.4    1120 35.6    1100 33.6    1032 63.6
      987 67.9       971 69.2     915 79.0     723 52.2     585 89.1
      527 93.1       495 90.3     473 86.7     463 86.6
END  29 PEAKS FOUND
```

TABLE 2 : Computer based interpretation for "unknown" oil.

```
POSSIBLE STRUCTURAL UNITS :       UNOIL

  202   ALKYL GROUP - GENERAL
  204   ALKYL GROUP - LONG CHAIN SUBSTITUENT
 1401   CARBOXYLIC ACID ESTER - POSSIBLY ALIPHATIC
 1417   LONG CHAIN ALIPHATIC ESTER - POSSIBLY UNSATURATED
 1418   LONG CHAIN ALIPHATIC ESTER - POSSIBLY UNSATURATED
 4905   CARBONYL COMPOUND - CLASS 5 (CONSULT MANUAL)
 4925   CARBONYL COMPOUND - POSSIBLY ESTER OR KETONE
```

TABLE 3 : Search results for "unknown" oil.

```
NEAREST 15 MATCHES IN LIBRARY :         UNOIL   SRCH1    05 S2

9-9  LU049A SOYBEAN OIL
9-8  LU049B SOYBEAN OIL
9-8  LU050A COTTONSEED OIL
9-8  LU051A PEANUT OIL; GROUND NUT OIL
9-8  LU053A LARD OIL
9-7  LU050B PEANUT OIL; GROUND NUT OIL
9-7  LU052A MUSTARD SEED OIL
9-6  LU049B COTTONSEED OIL
9-6  LU051B WALNUT OIL
9-6  LU055B LINSEED OIL
9-6  LU056A SPERM WHALE OIL
9-6  LU157B FATTY ESTER/ACID BASED RUST INHIBITOR
9-6  LU209B SULFONATED VEGETABLE OIL
9-5  LU054A BLOWN FISH OIL
9-5  LU056B SPERM WHALE OIL
```

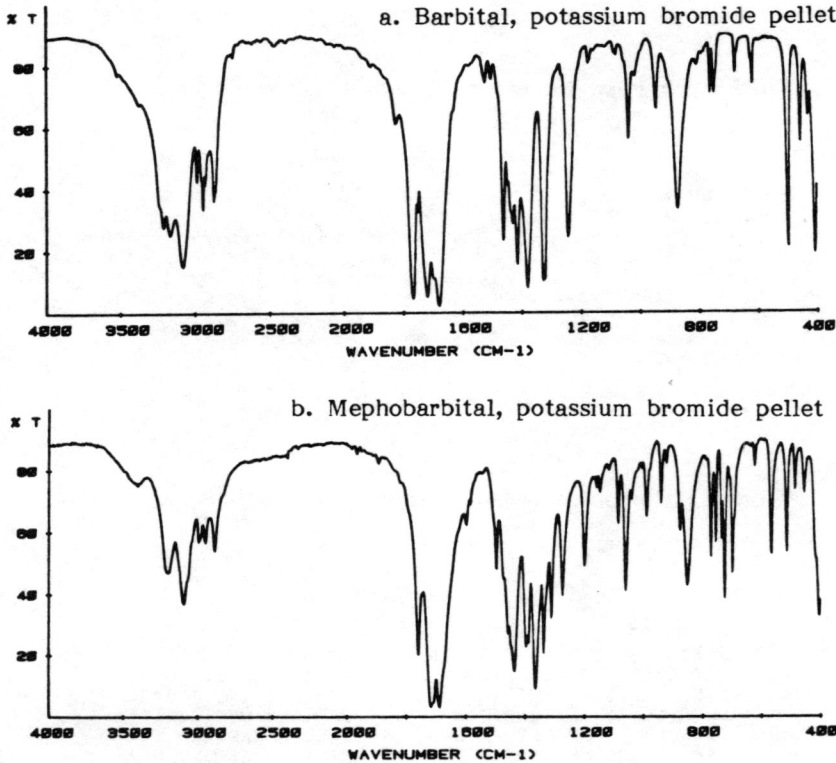

FIGURE 4 : The infrared spectra of a) barbital and b) mephobarbital.

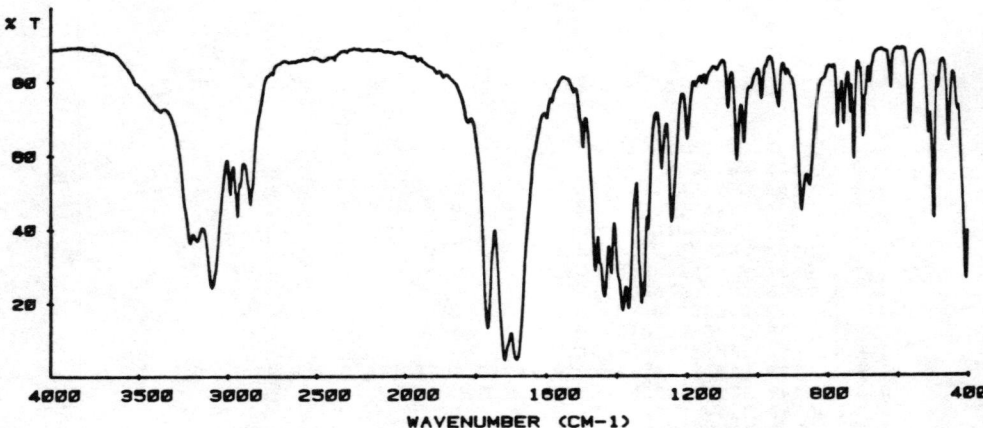

FIGURE 5 : A 1:1 mixture of barbital and Mephobarbital, potassium bromide pellet.

barbiturate, also shown in Table 4, has three carbonyl groups in widely different molecular environments. These are accounted for by the three different classes of carbonyl compound indicated in the computer output. The other interpretive statements also accurately describe the functionality of the sample.

At this point the spectral data can be searched by one of two schemes - either an absolute match or a positive match. The absolute match is intended for cases where the sample is to be considered as a single entity. In this case there is a negative weighting applied if bands are absent in either the library spectrum or the sample spectrum. The positive match scheme only counts bands that match and this is usually more suitable for mixture analysis. These two schemes differ in terms of their ability to discriminate in the searching process as demonstrated by Figures 6 and 7.

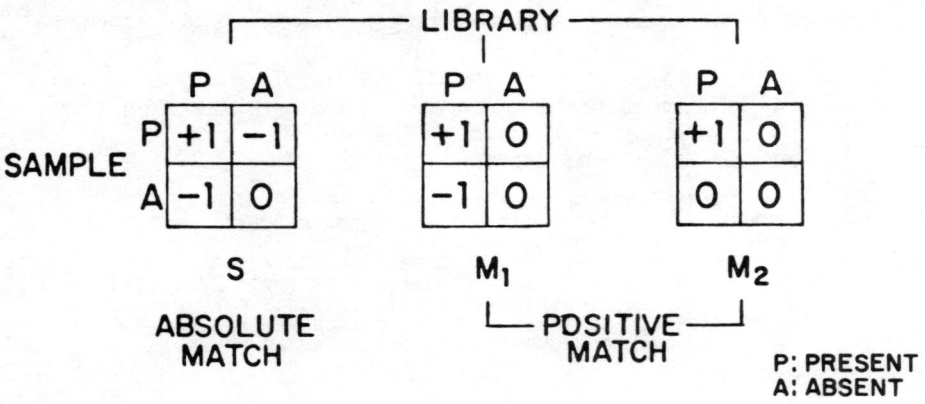

FIGURE 6 : Simple truth tables for single component and mixture based schemes.

Figure 6 provides simple truth tables that demonstrate the main differences between the search strategies. The relative scores for the small spectral segments in Figure 7 illustrate the discrimination for the S, M1 and M2 criteria from Figure 6. In this example the left hand spectrum is treated as the unknown and it is compared to itself and the other two spectra. In the case of a perfect match all three match criteria give the same score. A comparison of the unknown to the center spectrum gives the greatest difference for the absolute match (S); M1 and M2 have identical scores. A large difference is detected for the absolute match with the final spectrum. This time M1 and M2 give different scores indicating that M1 is more discriminating, where a negative weighting is applied to the broad band in the library spectrum at 1450 cm-1. This is a simplification of peak search strategies and in practical systems additional weightings are normally applied to account for the total number of peaks in the unknown and the library spectra.

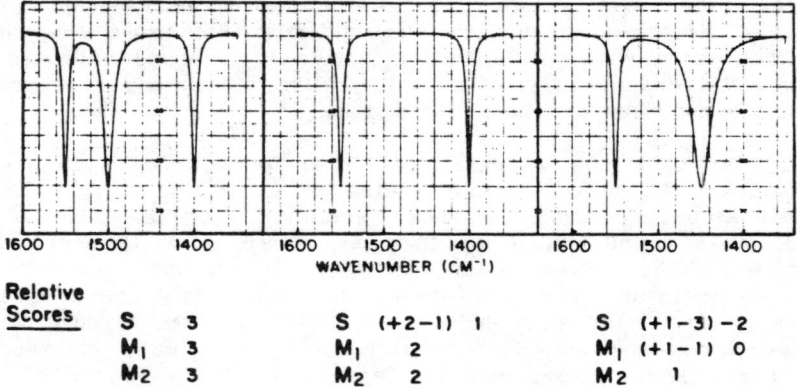

Relative Scores							
S	3	S	(+2−1) 1	S	(+1−3)	−2
M₁	3	M₁	2	M₁	(+1−1)	0
M₂	3	M₂	2	M₂		1

FIGURE 7 : Comparison of single component and mixture schemes.

TABLE 4 : Computer based interpretation for barbiturate mixture.

```
POSSIBLE STRUCTURAL UNITS :      DRUG1

 201   ALKYL GROUP - GENERAL
 203   ALKYL GROUP - POSSIBLY SHORT CHAIN SUBSTITUENT
1719   ALIPHATIC AMIDE - CYCLIC
4802   CARBONYL COMPOUND - POSSIBLY MULTICENTRED
4807   CARBONYL COMPOUND - POSSIBLY BARBITURIC ACID TYPE
4911   CARBONYL COMPOUND - CLASS 11 (CONSULT MANUAL)
4912   CARBONYL COMPOUND - CLASS 12 (CONSULT MANUAL)
4917   CARBONYL COMPOUND - CLASS 17 (CONSULT MANUAL)
```

Structure fragment for Barbiturate
(R, R' and R" vary for different barbiturates)

Returning to the original problem of the binary drug mixture, we can clearly benefit by adopting an M1 type mixture strategy. The results of this type of search are given in Table 5. The two components of the mixture are given the highest ratings. The peak match scores are not perfect because of overlap. In this example the interpretation was given a high weighting and this biassed the results to compounds of similar structure as evidenced by the other compounds given in the list. This feature is particularly useful when characterizing mixtures of similar compounds. Mixtures of dissimilar compounds are best analyzed with reduced weighting from the interpretation.

TABLE 5 : Mixture search results for 1:1 barbiturate mixture.

```
NEAREST 15 MATCHES IN LIBRARY :         DRUG1           05 S2

    9-6 PE127B MEPHOBARBITAL
    9-4 PE127C BARBITAL - DIETHYLBARBITURIC ACID
    9-2 PE129B PENTOBARBITAL
    9-2 PE134B PHENOBARBITAL
    9-2 PE135A BREVITAL METHOHEXITAL SODIUM - CAST FILM
    9-1 CD391A BENZO(G)PTERIDINE-2,4(1H,3H)-DIONE
    9-1 PE500A AMOBARBITAL - COMPOUND
    9-1 PE126A AMOBARBITAL
    9-1 CD325A 1,3-DIAZASPIRO(4,4)NONANE-2,4-DIONE
    9-1 PE130B BUTABARBITAL
    9-1 PE128B DIALLYLBARBITURIC ACID
    9-0 PE126B BUTETHAL
    9-0 PE124B BREVITAL - METHOHEXITAL SODIUM
    9-0 PE127A BARBITAL - COMPOUND
    9-0 PE131A SECOBARBITAL
```

It must be emphasized at this point that this approach to mixture analysis can have many pitfalls. A few of the major problem areas are listed below :

    a) key bands may be masked by overlap - as partially experienced in the example above,
    b) band intensities and/or positions can be distorted by overlap,
    c) band intensities and/or positions can be distorted by mutual interaction,
    d) hydrogen bonding can distort or mask absorptions,
    e) accidental coincidences can lead to mismatches,
    f) loss of crystallinity in mixtures can reduce scores for certain solid samples.

Any search method applied to mixtures will work for certain selected examples, even for some complex mixtures where a carefully controlled, small library is used. For a true unknown, however, it would be foolish to rely solely on a search output. Some additional spectral manipulation or chromatographic separation would be necessary to confirm the results.

Many commercial products are complex chemical or physical mixtures. In some cases the use of the type of search scheme described above is not always useful. Sometimes, all that is required is a broad based characterization or a generic identity in terms of a product type. A good example is a polychlorinated biphenyl (a PCB), such as the Arochlor 1254 shown in Figure 8. This time the identification of the individual component polychlorinated compounds would not be useful. Therefore in this case a normal absolute scoring scheme is preferred where the material is treated as a single entity. The computer interpretation (Table 6) accurately classifies the sample as an aromatic material with multiple halogen (ring) substituents.

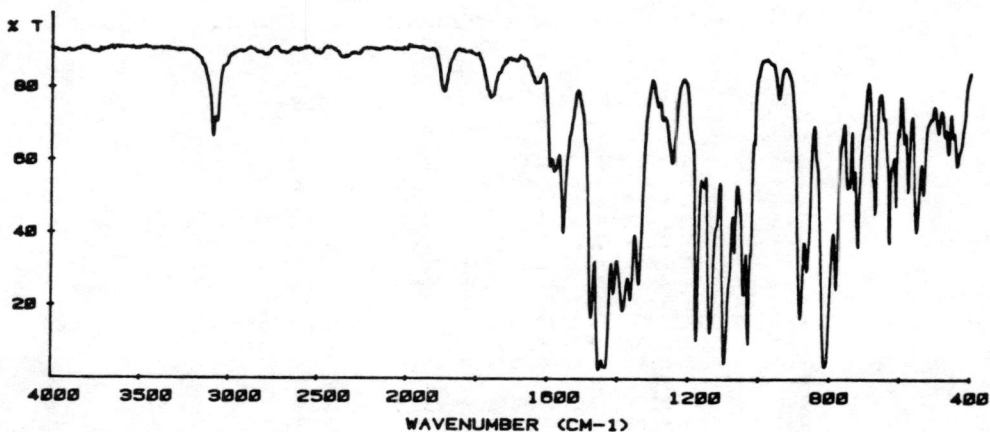

FIGURE 8 : Polychlorinated biphenyl, Arochlor 1254, thin film.

TABLE 6 : Computer based interpretation for Arochlor 1254.

```
POSSIBLE STRUCTURAL UNITS :        UNK01

  254   AROMATIC OR UNSATURATED COMPOUND
  257   AROMATIC COMPOUND - GENERAL
  269   PHENYL OR SUBSTITUTED ARYL COMPOUND - GENERAL
  276   PARA ARYL, ARYLOXY OR ARYLAMINO COMPOUND
 2211   AROMATIC COMPOUND - POSSIBLY MULTIPLE HALOGEN SUBSTITUTION
```

TABLE 7 : Search results for Arochlor 1254.

```
NEAREST 15 MATCHES IN LIBRARY :           UNK01   SRCH1    05 S2

   9-9   PP007A  AROCHLOR 1254
   9-6   PP006B  AROCHLOR 1248
   9-4   PP004B  AROCHLOR 1016
   9-4   PP006A  AROCHLOR 1242
   9-3   PP005B  AROCHLOR 1232
   9-2   CH097B  BENZENE, 1-(TRIFLUOROMETHYL)-2-IODO-
   9-2   PE029A  1,2,4-TRICHLOROBENZENE
   9-2   CH085B  BENZENE, 1-BROMO-2,5-DICHLORO-
   9-2   SC485B  1,2,3-TRICHLOROBENZENE
   9-1   CH060A  BENZENE, 1,2,4-TRICHLORO-
   9-1   CH085A  1-BROMO-3,4-DICHLORO-
   9-1   CH056A  BENZENE, 1-(BROMOMETHYL)-2-METHYL-
   9-1   PP005A  AROCHLOR 1221
   9-1   CH090A  BENZENE, 1-BROMO-2-FLUORO-
   9-1   CH074A  BENZENE,1-CHLORO-2-(2,2,2-TRICHLORO-1-
                          (4-CHLOROPHENYL)ETHYL)-
```

An exact match is obtained in the search for the Arochlor 1254 and interestingly, the four closest matches, with significantly reduced scores, are also Arochlors (Table 7). All the other compounds in the list are halogen substituted aromatic hydrocarbons. The key point in this example is that priority has been given to the materials of common generic type.

7. CONCLUSIONS

The purpose of this article has been to demonstrate the practical application of computer based search systems to common analytical problems. In many cases it is possible for the computer to perform the same role as the spectroscopist. This is achieved by the combination of a computer based interpretation with a flexible search program. The main advantages of this type of system are :

    a) Functional group assignments can be made irrespective of the level of expertise of the operator.

    b) A pre-sort can be made to improve the selectivity of a search.

    c) A broad based characterization can often be obtained without a library.

    d) Compounds of similar structure or functionality are usually grouped together. This aids identification if the sample spectrum is not in the library.

REFERENCES

1. Sparks,R.A.,"Storage and Retrieval of Wyandotte-ASTM IR spectral Data using an IBM 1401 Computer", ASTM, Philadelphia, 1964.

2. Anderson,D.H. and Covert,G.L., Anal. Chem., 39,(1967),p.1288.

3. Erley,D.S., Anal. Chem., 40,(1968),p.894.

4. Erley,D.S., Appl. Spectrosc., 25,(1971),p.200.

5. ASTM method E 204-78,"Identification of Material by Infrared Absorption Spectroscopy, Using the ASTM Coded Band and Chemical Classification Index, ASTM, 1916 Race Street, Philadelphia, Pennsylvania 19103.

6. Kuentzel,L.E., Anal. Chem., 23,(1951),p.1413.

7. IRGO is a spectral search system that is offered by : Chemir Laboratories, 761 West Kirkham, Glendale, Missouri 63122, USA.

8. Rann,C.S., Anal. Chem., 44,(1972),p.1669.

9. Hippe,Z. and Hippe,R.,Appl. Spectrosc. Rev., 16(1),(1980), pp.135-186.

10. Ford,M.A., Carter,H., White,P.P., Coates,J.P., Savitzky,A., Geary,S., Muir,A. and Hannah,R.W., Pittsburgh Conference on Analytical Chemistry and Applied Spectroscopy, Cleveland, March 5-9, 1979.

11. Leupold,W-R., Domingo,C., Niggemann,W. and Schrader,B., Fresenius' Z. Anal. Chem., 303, (1980), p.337.

12. Visser,T. and van der Maas,J.H., Anal. Chim. Acta, 122, (1980), p.357.

13. Visser,T. and van der Maas,J.H., Anal. Chim. Acta, 133, (1981), p.451.

14. Farkas, M., Markos,J., Szepesvary,P., Bartha,I., Szalontai,G. and Simon,Z., Anal. Chim. Acta, 133, (1981), p.19.

15. Tomellini,S.A., Saperstein,D.D., Stevenson,J.M., Smith,G.M., Woodruff,H.B. and Seelig,P.F., Anal. Chem., 53,(1981), p.2367.

16. Kwiatkowski,J. and Riepe,W., Anal. Chim. Acta, 135, (1982), p.285.

17. Kwiatkowski,J. and Riepe,W., Anal. Chim. Acta, 135, (1982), p.293.

18. Hippe,Z., Anal. Chim. Acta, 150, (1983), p.11.

19. Gribov,L.A., Elyashberg,M.E., Koldashov,V.N. and Pletnjov,I.V. Anal. Chim. Acta, 148,(1983), p.159.

20. Smith,G.H. and Woodruff,H.B., J. Chem. Inf. Comput. Sci., 24, (1984), p.33.

21. "Spec-Finder" is a registered trademark of the Sadtler Research Laboratories, Division of Bio-Rad Inc., 3316 Spring Garden Street, Philadelphia, Pennsylvania 19104, USA.

22. Hummel,D.O.,"Atlas of Polymer and Plastics Analysis", Volume 1, "Polymers:Structures and Spectra", Verlag Chemie, Weinheim - Deerfield Beach, Florida - Basel.

23. Mills,T III, Price, W.N., Price, P.T. and Roberson,J.C.,"Instrumental Data for Drug Analysis", Volume 1, (1982), Elsevier, New York - Amsterdam - Oxford.

24. Mills,T III, Price, W.N. and Roberson, J.C., "Instrumental Data for Drug Analysis", Volume 2, (1984), Elsevier, New York - Amsterdam - Oxford.

25. Coates,J.P. and Setti,L.C.,"Oils, Lubricants and Petroleum Products", (1983), Perkin-Elmer Corporation, Main Avenue, Norwalk, Connecticut, USA.

26. Chicago Society for Coatings Technology, "An Infrared Spectroscopy Atlas for the Coatings Industry", Federation of Societies for Coatings Technology, 1315 Walnut Street, Philadelphia, Pennsylvania 19107, USA.

AUTHOR INDEX

Alix, A.J.P., 3
Anderson, R.J., 60

Basile, J.R., 139
Beer's Law, 60
Born-Oppenheimer, 36
Buijs, H., 43
Butler, I.S., 83

Cameron, D.G., 159
Chase, D.B., 70
Chenery, D.H., 20
Chesters, M., 97
Claiborne, R., 126
Coates, J.P., 167
Coblentz, W.W., 26
Connes, P., 41,50
Cooley-Tukey algorithm, 36,45
Crawford's rule, 7

Eckart conditions, 7
Einstein's coefficients, 6
Erickson, M.D., 66

Fateley, W.G., 25
Fellgett, P.G., 20,36,46
Ferrerro, J.R., 139
Fourier Transform, 35,44,45

Gersho, A., 154
Grasselli, J.G., 25,55
Gray, H.B., 83
Griffith, P.R., 60,70,153
Grim, W.M., III,25

Haaland, D.H., 74
Hannah, R.W., 167

Ismail, A.A., 83

Jacquinot, P., 20,41,46
Jakobsen, R.J., 74

Kauppinen J.K., 75
Kettle, S.F.A., 17,97
Kinney, J.B., 72
Krishnan, K., 139,154,157
Kubelka-Munk, 151,152
Kuehl, D., 70

Langrange equations, 5
Laux, L., 20

Lorentz-Gauss, 165

Mamantov, G., 72
Mantsch, H., 125
Mantz, A.W., 74
Martin and Puplett, 49,50
Michelson, A.A., 35,38,153
Miknis, F.P., 60

Nafie, L.A., 50
Nakamoto, K., 83

Overend, J., 20
Oxton, I., 97

Placzek's approximation, 6
Planck, M., 28
Powell, D., 97
Powel, G.L., 70

Riseman, S.M., 72
Rosencwaig, A., 154

Savitzky-Goley, 165
Scanlon, K., 20
Sedman, J., 83
Sheppard, N., 20,97
Singer and Nicolson's model, 126
Smyrl, N.R., 153

Solomon, P.R., 66
Stanley, R.H., 72

Teller-Redlich product rule, 5
Theophanides, T., 105

Vidrine, D.W., 154

Watson-Crick, 106
Weissmann, G., 126
Wilson formalism, 5

SUBJECT INDEX

Ab initio, 18
Absolute infrared intensities, 3ff
Additivity rule, 9
Adenine, 107
Adenosine, 107
Adenosine diphosphate (ADP), 110
Adenosine monophosphate, (AMP), 107
Adenosine triphosphate (ATP), 110
Advantages of Fourier Transform IR, 58
AgBr/AgCl windows, 28
Amplitude of vibrations, 3ff
Apodize, 20
Apodization, 59
Application of FT-IR, 103
Arochlor, 181,182
Aspirin, 156
Attenuated total reflectance (ATR), 26,56

Ba^{++}, 117
Background artifacts, 172
Background correction, 58
Band fitting, 162
Banwidth measurements, 159
Barbital, 178,181
Beamsplitters, 40
Biomedical pharmaceutical, 56
Biomembranes, 132
Black-body, 28,30,55
Blood, 84
Bolometer, 33
Born-Oppenheimer approximations, 3

Ca^{++}, 110,117
Caffeine, 152

Calf thymus DNA, 119
Carbonyl clusters, 101
$C-CH_3$, 100
$C-CD_3$, 100
$\nu C=C$, 105
$\nu C-C$, 115
νCH_2, 100,105
νCH, 105,171
$\nu C=N$, 105
Chopper, 29
$C=O$ stretching, 105
$\nu(CO)$, 95,96
Coal, 65
Complexed protein spectra, 92
Computed assisted spectral interpretation, 167,174
Computerized IR instruments, 56
Connes advantages, 41
Coriolis coupling constant, 3,7
Cooley-Tukey algorithm, 36
Coordination compounds, 83
Copper, 119
Cosine wave, 40
$Cr(CO)_6$, 85
Crawford's rule, 7
CsI/CsBr windows, 28
Cytidine monophosphate (CMP),107
Cytidine, 107
Cytosine, 107
Cytosol, 84

DNA, B-form, 115
Data collection, 159
Data processing in FT-IR, 56,159

Deconvolution, 126,162
Deoxyribonucleic acid (DNA), 106
Derivation, 162
Detectors, 29,33,48
Diamond windows, 26,28
Difference spectra, 61,162
Diffractive grating spectrometer, 44
Diffuse reflectance, 26,148,150
Dimethyl propylene-urea, 83
Dipole moment derivatives, 7,9
Dispersive Infrared Spectroscopy, 25,27
Dispersive instruments, 20,25
DNA-histones, 110
Doppler broadening, 21
Drug mixture, 180
DTGS detector, 33

Electrical filters, 56
Encapsulation in KBr pellets, 26
Endrin, 151
Estrogen, 84
Evolved gas analysis/FT-IR, 56

Factor analysis, 165
False energy, 32
Far IR detectors, 48
Felgett, and Jacquinot advantage, 19,46
FIR (Far Infrared), 28,35,40,48
Fourier transform spectroscopy, 43
Frequencies, 112,159
FT-IR/ATR accessory, 142
FT-IR difference spectrum, 63
FT-IR/PAS, 155
FT-IR spectra, 36,105
FT-spectroscopy, 47

5'-GMP, 113,118
Gas chromatogram, 68
Gases, 26
GC/FT-IR, 56,66
GC-IR, 57
Genzel Interferometer, 39
G-F Wilson formalism, 5
Globar, 29
Gratings, 27,29,31
Guanine, 107
Guanosine, 107
Guanosine monophosphate (GMP), 107

Heavy hydrocarbons, 65
Helium-Neon laser, 45
Hg-Arc, 29
Hg-Cd-Te (MCT) detector, 33
High resolution spectroscopy, 50
Hormone-receptor, 84
Hydrogen bonds, 107
Hypoxanthine, 107

Identification of a single
    unknown, 169,175
Industrial applications of
    FT-IR, 55,65
Infrared intensities, 3,6,17
Inosine, 107
Inosine monophosphate (IMP), 107
InSb detector, 33
Integrated absorption
    intensities, 7
Interferogram, 39
Interferometer, 50
Interferometric infrared
    spectroscopy, 25,27
Internal reflectance, 37

SUBJECT INDEX

Interpretive search system, 167,169
Intensities of infrared spectral bands, 17
Isotopic intensity shift, 3,10

KBr disk, 28,178
Key bands, 181
Kinetic Energy, 3
Kubelca-Munk functions, 152

LC/FT-IR, 56
Liquid chromatography, 70
Liquids, 26
Lube oil Analysis, 60
Lyophylized samples, 126
Lysolecithins, 126,128

Materials used as windows, 28
Matrix isolation, 56
Membrane metastability, 130
Membranes, 125
Mephobarbital, 178,181
Metal clusters, 97
Metal complexes, FT-IR spectra, 114
Metal ions, 105,110
Metal surfaces, 97
Mg^{++}, 110,117,119
$Mg(H_2O)_5(GMP)^{++}$ complex ion, 114
$\nu(M-H)$, 100
Michelson Interferogram, 39
Mid-IR, 36,40
Minerals, 65
Mobile FT-IR, 70,71
Molecular constants, 7
Molecular vibrations, 3
Monochromatic frequency, 19
Multicomponent analysis, 165
Mylar beamsplitter, 48,49

NaCl, 28
Near infrared, 40
Nerst glower, 29
νNH, 105
$Ni(H_2O)_5(GMP)^{++}$ complex ion, 114
NIR (near infrared), 28,35,36
Nomenclature, 106
Normal coordinate analysis, 17
Nucleic acids, 105
Nucleoproteins, 110
Nujol mull, 26

νOH, 105
O-P-O stretch, 115

PAS/FT-IR, 72
PAS, photoaconstic spectrocopy, 56,147,152
PbS detector, 33
PbSe detector, 33
Periodic table, 111
Phenacetin, 156
Phosphate frequencies, 105,117
Phospholipids, 128
Photoacoustic detection, 72
PO_2^- stretch, 115
Polyetylene, 28
Polymers, 62
Polysteryne spectra, 59
Precision and reproducibility, 159
Product rule, 12
Cis-$[Pt(NH_3)_2(GMP)_2]Cl_2$, 114,116
Purines, 105
Pyrimidines, 105

Quartz, 28

Receptor, 84
Reflection, 261
Reproducibility, 46
Ribonucleic Acid (RNA), 108
Ribosome, 110
RNA-protein, 110

Sample handling techniques, 26,112
Sample thickness, 173
Sampling Accessories, 57
Search system, 172
Self compensation, 51
Sensitivity, 47
Shale, 65
Signal to noise, 20,41,47
Silicon, 26
Single strand, 109
Site symmetry, 101
Smoothing, 162,165
ν Sn-Br
ν Sn-O
ν Sn-Ph
Solids, 26
Software, 171,175
Software normalization, 171
Spectral search, 174
Spectrum normalization, 172
Symmetry related internal coordinates, 18
Symmetry species, 17

Teller-Reddlich product rule, 12
Theory of Molecular vibrations, 3
Thermotropic mesomorphism, 126
TGS detector, 33
Thin film sample, 182
Thymidine, 107
Thymidine monophosphate (TMP), 107

Thymine, 107
Trace-Analysis, 73
Triphenyltin (IV) bromide, 83

Uracil, 107
Uridine, 107
Uridine monophosphate (UMP), 107

Watson-Crick pairing, 106
Wavelength, 30,38
Wavenumber, 28,38
Width, 161
W-Lamp, 29

Zeolite, 157
Zero Crossing, 40